MECHANICAL ENGINEERING
PROBLEMS & SOLUTIONS

Seventh Edition

Jerry H. Hamelink, PhD, PE & John D. Constance, PE

President: Roy Lipner
Vice-President of Product Development and Publishing: Evan M. Butterfield
Editorial Project Manager: Laurie McGuire
Director of Production: Daniel Frey
Creative Director: Lucy Jenkins

Published by Kaplan® AEC Education,
a division of Dearborn Financial Publishing, Inc.®,
a Kaplan Professional Company®

30 South Wacker Drive
Chicago, IL 60606-7481
(312) 836-4400
www.kaplanaecengineering.com

CONTENTS

Introduction

HOW TO USE THIS BOOK

Mechanical Engineering: Problems & Solutions and its companion text, *Mechanical Engineering: License Review*, form a two-part approach to preparing for the Principles and Practice of Mechanical Engineering exam:

Mechanical Engineering License Review contains the conceptual review of mechanical engineering topics for the exam. Solved examples illustrate how to apply the equations and analytical methods discussed in the text.

Mechanical Engineering Problems and Solutions provides problems for you to solve in order to test your understanding of concepts and techniques. Ideally, you should solve these problems after completing your conceptual review. Then, compare your solution to the detailed solutions provided, to get a sense of how well you have mastered the content and what topics you may want to review further.

BECOMING A PROFESSIONAL ENGINEER

To achieve registration as a professional engineer there are four distinct steps: (1) education, (2) the Fundamentals of Engineering/Engineer-In-Training (FE/EIT) exam, (3) professional experience, and (4) the professional engineer (PE) exam, more formally known as the Principles and Practice of Engineering Exam. These steps are described in the following sections.

Education

The obvious appropriate education is a B.S. degree in Mechanical engineering from an accredited college or university. This is not an absolute requirement. Alternative, but less acceptable, education is a B.S. degree in something other than mechanical engineering, or a degree from a non-accredited institution, or four years of education but no degree.

Fundamentals of Engineering (FE/EIT) Exam

Most people are required to take and pass this eight-hour multiple-choice examination. Different states call it by different names (Fundamentals of Engineering, E.I.T., or Intern Engineer), but the exam is the same in all states. It is prepared and graded by the National Council of Examiners for Engineering and Surveying (NCEES). Review materials for this exam are found in other Kaplan AEC books such as *Fundamentals of Engineering: FE Exam Preparation*.

Experience

Typically one must have four years of acceptable experience before being permitted to take the Professional Engineer exam (California requires only two years). Both the length and character of the experience will be examined. It may, of course, take more than four years to acquire four years of acceptable experience.

Professional Engineer Exam

The second national exam is called Principles and Practice of Engineering by NCEES, but just about everyone else calls it the Professional Engineer or P.E. exam. All states, plus Guam, the District of Columbia, and Puerto Rico, use the same NCEES exam.

MECHANICAL ENGINEERING PROFESSIONAL ENGINEER EXAM

The reason for passing laws regulating the practice of Mechanical engineering is to protect the public from incompetent practitioners. Most states require engineers working on projects involving public safety to be registered, or to work under the supervision of a registered engineer. In addition, many private companies encourage or require engineers in their employ to pursue registration as a matter of professional development. Engineers in private practice, who wish to consult or serve as expert witnesses, typically also must be registered. There is no national registration law; registration is based on individual state laws and is administered by boards of registration in each of the states. A listing of the state boards is in Table 1.1.

Examination Development

Initially the states wrote their own examinations, but beginning in 1966 the NCEES took over the task for some of the states. Now the NCEES exams are used by all states. This greatly eases the ability of an engineer to move from one state to another and achieve registration in the new state.

Table 1.1 State boards of registration for engineers

State/Territory	Web site	Telephone
AK	www.dccd.state.ak.us/occ/pael.htm	(907) 465-1676
AL	www.bels.state.al.us	(334) 242-5568
AR	www.state.ar.us/pels	(501) 682-2824
AZ	www.state.ar.us/pels	(602) 364-4930
CA	dca.ca.gov/pels/contacts/htm	(916) 263-2230
CO	dova.state.co.us/engineers _surveyors	(303) 894-7788
CT	state.ct.us/dcp	(806) 713-6145
DC		(202) 442-4320
DE	www.dape.org	(302) 368-6708
FL	www.fbpe.org	(850) 521-0500
GA	www.sos.state.ga.us/plb/pels/	(478) 207-1450
GU	www.guam-peals.org	(671) 646-3138
HI	www.hawaii.gov/dcca/pbl	(808) 586-2702
IA	www.ia.us/government/com	(515) 281-4126
ID	www.state.id.us/ipels/index.htm	(208) 334-3860
IL	www.kpr.state.il.us	(217) 785-0877
IN	www.in.gov/pla/bandc/engineers	(317) 232-2980
KS	www.accesskansas.org/ksbtp	(785) 296-3053
KY	www.kybocls.state.ky.us	(502) 573-2680
LA	www.lapels.com	(225) 925-6291
MA	www.state.ma.us/reg	(617) 727-9957
MD	www.dllr.state.md.us	(410) 230-6322
ME	www.professionals.maincusa.com	(207) 287-3236
MI	www.michigan.gov/cis/0,1607.7-154-10557_12992_14016—.00.htm.us	(517) 241-9253
MN	www.aclslagid.state.mn.us	(651) 296-2388
MO	www.pr.mo.gov/apelsla.asp	(573) 751-0047
MP		(011)-(670) 234-5897
MS	www.pepls.state.ms.us	(601) 359-6160
MT	www.discoveringmontana.com/dli/bsb/license/bsd_board/pel_board/board_page.htm	(406) 841-2367
NC	www.ncbels.org	(919) 881-4000
ND	www.ndpelsboard.org/	(701) 258-0786
NE	www.ea.state.ne.us	(402) 471-2021
NH	www.state.nh.us/jtboard/home.htm	(603) 271-2219
NJ	www.state.nj.us	(973) 504-6460
NM	www.state.nm.us/pepsboard	(518) 827-7561
NV	www.boe.state.nv.us	(775) 688-1231
NY	www.op.nysed.gov	(518) 474-3846 3817×140
OH	www.ohiopeps.org	(614) 466-3651
OK	www.pels.state.ok.us/	(405) 521-2874
OR	www.osbeels.org	(503) 362-2666
PA	www.dos.state.pa.us/eng	(717) 783-7049
PR	P.O. Box 3271, San Juan 00904	(787) 722-2122
RI	www.bdp.state.ri.us	(401) 222-2565
SC	www.lln.state.sc.us/POL/Engineers	(803) 896-4422

(Continued)

Table 1.1 State boards of registration for engineers *(Continued)*

State/Territory	Web site	Telephone
SD	www.state.sd.us/dol/boards/engineer	(605) 394-2510
TN	www.state.tn.us/commerce/boards/ae	(615) 741-3221
TX	www.tbpe.state.tx.us	(512) 440-7723
UT	www.dopl.utah.gov	(801) 530-6632
VA	www.state.va.us/dopr	(804) 367-8514
VI	www.dlca.gov.vi/pro-aels.html	(340) 773-2226
VT	www.vtprofessionals.org	(802) 828-3256
WA	www.dol.wa.gov/engineers/engfront.htm	(360) 664-1595
WI	www.drl.state.wi.us	(608) 261-7096
WV	www.wvpebd.org	(304) 558-3554
WY	www.wrds.uwyo.edu/wrds/borpe/borpe.html	(307) 777-6155

The development of the engineering exams is the responsibility of the NCEES Committee on Examinations for Professional Engineers. The committee is composed of people from industry, consulting, and education, plus consultants and subject matter experts. The starting point for the exam is a task analysis survey, which NCEES does at roughly 5- to 10-year intervals. People in industry, consulting, and education are surveyed to determine what mechanical engineers do and what knowledge is needed. From this NCEES develops what it calls a "matrix of knowledge" that forms the basis for the exam structure described in the next section.

The actual exam questions are prepared by the NCEES committee members, subject matter experts, and other volunteers. All people participating must hold professional registration. Using workshop meetings and correspondence by mail, the questions are written and circulated for review. The problems relate to current professional situations. They are structured to quickly orient one to the requirements, so that the examinee can judge whether he or she can successfully solve it. Although based on an understanding of engineering fundamentals, the problems require the application of practical professional judgment and insight.

Examination Structure

The exam is organized into breadth and depth sections.

The morning breadth exam consists of 40 multiple-choice questions covering the following areas of mechanical engineering (relative exam weight for each topic is shown in parentheses):

- General knowledge, codes and standards (30%)

- Machine design and materials (17%)

- Hydraulics and fluids (17%)

- Energy conversion and power systems (18%)

- HVAC and refrigeration (18%)

You will have four hours to complete the breadth exam.

The afternoon depth exam is actually three exams; you choose the depth exam you wish to take. The depth exams are

- HVAC and Refrigeration

- Machine Design

- Thermal and Fluids Systems

Clearly, you should choose the exam that best matches your training and professional practice. You will have four hours to answer the 40 multiple-choice questions that make up the depth exam.

Both the breadth and depth questions include four possible answers (A, B, C, D) and are objectively scored by computer.

For more information on the topics and subtopics and their relative weights on the breadth and depth portions, visit the NCEES Web site at www.ncees.org.

Exam Dates

The National Council of Examiners for Engineering and Surveying (NCEES) prepares Mechanical Engineering Professional Engineer exams for use on a Friday in April and October of each year. Some state boards administer the exam twice a year in their state, whereas others offer the exam once a year. The scheduled exam dates are:

	April	October
2005	15	28
2006	21	27
2007	20	26
2008	11	24

People seeking to take a particular exam must apply to the state board several months in advance.

Exam Procedure

Before the morning four-hour session begins, the proctors pass out an exam booklet and solutions pamphlet to each examinee.

The solution pamphlet contains grid sheets on right-hand pages. Only the work on these grid sheets will be graded. The left-hand pages are blank and are to be used for scratch paper. The scratchwork will not be considered in the scoring.

If you finish more than 30 minutes early, you may turn in the booklets and leave. In the last 30 minutes, however, you must remain to the end to ensure a quiet environment for all those still working and the orderly collection of materials.

The afternoon session will begin following a one-hour lunch break. The afternoon exam booklet will be distributed along with an answer sheet.

Exam-Taking Suggestions

People familiar with the psychology of exam taking have several suggestions for people as they prepare to take an exam.

1. Exam taking involves really, two skills. One is the skill of illustrating knowledge that you know. The other is the skill of exam taking. The first may be enhanced by a systematic review of the technical material. Exam-taking skills, on the other hand, may be improved by practice with similar problems presented in the exam format.

2. Since there is no deduction for guessing on the multiple choice problems, answers should be given for all of them. Even when one is going to guess, a logical approach is to attempt to first eliminate one or two of the four alternatives. If this can be done, the chance of selecting a correct answer obviously improves from 1 in 4 to 1 in 3 or 1 in 2.

3. Plan ahead with a strategy. Which is your strongest area? Can you expect to see several problems in this area? What about your second strongest area? What will you do if you still must find problems in other areas?

4. Plan ahead with a time allocation. Compute how much time you will allow for each of the subject areas in the breadth exam and the relevant topics in the depth exam. You might allocate a little less time per problem for those areas in which you are most proficient, leaving a little more time in subjects that are difficult for you. Your time plan should include a reserve block for especially difficult problems, for checking your scoring sheet, and to make last-minute guesses on problems you did not work. Your strategy might also include time allotments for two passes through the exam—the first to work all problems for which answers are obvious to you, and the second to return to the more complex, time-consuming problems and the ones at which you might need to guess. A time plan gives you the confidence of being in control and keeps you from making the serious mistake of misallocation of time in the exam.

5. Read all four multiple-choice answers before making a selection. An answer in a multiple-choice question is sometimes a plausible decoy—not the best answer.

6. Do not change an answer unless you are absolutely certain you have made a mistake. Your first reaction is likely to be correct.

7. Do not sit next to a friend, a window, or other potential distractions.

Exam Day Preparations

The exam day will be a stressful and tiring one. This will be no day to have unpleasant surprises. For this reason we suggest that an advance visit be made to the examination site. Try to determine such items as

1. How much time should I allow for travel to the exam on that day? Plan to arrive about 15 minutes early. That way you will have ample time, but not too much time. Arriving too early, and mingling with others who also are anxious, will increase your anxiety and nervousness.

2. Where will I park?

3. How does the exam site look? Will I have ample workspace? Where will I stack my reference materials? Will it be overly bright (sunglasses), cold (sweater), or noisy (earplugs)? Would a cushion make the chair more comfortable?

4. Where are the drinking fountain, lavatory facilities, pay phone?

5. What about food? Should I take something along for energy in the exam? A bag lunch during the break probably makes sense.

What to Take to the Exam

The NCEES guidelines say you may bring only the following reference materials and aids into the examination room for your personal use:

1. Handbooks and textbooks, including the applicable design standards.

2. Bound reference materials, provided the materials remain bound during the entire examination. The NCEES defines "bound" as books or materials fastened securely in their covers by fasteners that penetrate all papers. Examples are ring binders, spiral binders and notebooks, plastic snap binders, brads, screw posts, and so on.

3. Battery-operated, silent, nonprinting, noncommunicating calculators. Beginning with the April 2004 exam, NCEES has implemented a more stringent policy regarding permitted calculators. For more details, see the NCEES website (www.ncees.org), which includes the policy and a list of permitted calculators. You also need to determine whether or not your state permits preprogrammed calculators. Bring extra batteries for your calculator just in case; many people feel that bringing a second calculator is also a very good idea.

At one time NCEES had a rule barring "review publications directed principally toward sample questions and their solutions" in the exam room. This set the stage for restricting some kinds of publications from the exam. *State boards may adopt the NCEES guidelines, or adopt either more or less restrictive rules.* Thus an important step in preparing for the exam is to know what will—and will not—be permitted. We suggest that if possible you obtain a written copy of your state's policy for the specific exam you will be taking. Occasionally there has been confusion at individual examination sites, so a copy of the exact applicable policy will not only allow you to carefully and correctly prepare your materials, but will also ensure that the exam proctors will allow all proper materials that you bring to the exam.

As a general rule we recommend that you plan well in advance what books and materials you want to take to the exam. Then they should be obtained promptly so you use the same materials in your review that you will have in the exam.

License Review Books

The review books you use to prepare for the exam are good choices to bring to the exam itself. After weeks or months of studying, you will be very familiar with their organization and content, so you'll be able to quickly locate the material you want to reference during the exam. Keep in mind the caveat just discussed—some state boards will not permit you to bring in review books that consist largely of sample questions and answers.

Textbooks

If you still have your university textbooks, they are the ones you should use in the exam, unless they are too out of date. To a great extent the books will be like old friends with familiar notation.

Bound Reference Materials

The NCEES guidelines suggest that you can take any reference materials you wish, so long as you prepare them properly. You could, for example, prepare several volumes of bound reference materials, with each volume intended to cover a particular category of problem. Maybe the most efficient way to use this book would be to cut it up and insert portions of it in your individually prepared bound materials. Use tabs so that specific material can be located quickly. If you do a careful and systematic review of civil engineering, and prepare a lot of well-organized materials, you just may find that you are so well prepared that you will not have left anything of value at home.

Other Items

In addition to the reference materials just mentioned, you should consider bringing the following to the exam:

■ *Clock*—You must have a time plan and a clock or wristwatch.

■ *Exam assignment paperwork*—Take along the letter assigning you to the exam at the specified location. To prove you are the correct person, also bring something with your name and picture.

■ *Items suggested by advance visit*—If you visit the exam site, you probably will discover an item or two that you need to add to your list.

■ *Clothes*—Plan to wear comfortable clothes. You probably will do better if you are slightly cool.

■ *Box for everything*—You need to be able to carry all your materials to the exam and have them conveniently organized at your side. Probably a cardboard box is the answer.

CHAPTER

Mechanics
of Materials

OUTLINE

PROBLEMS

1.1 A projectile is fired at an angle of 37° from the horizontal at an initial velocity of 2000 fps across terrain as shown in Exhibit 1.1. Find the time it takes to reach point L on the plateau.

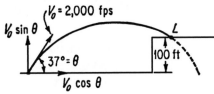

Exhibit 1.1

1.2 A sphere weighs 1500 lb. Find the tension in the wire and the direction and magnitude of the reaction at the hinge at A. Refer to Exhibit 1.2.

1.3 A rocket bomb weighs 14 tons, of which 6 are fuel plus oxygen. In order to simplify the calculations, suppose the rocket to be projected vertically upward and the stream of gas ejected to be at a constant rate of 260 lb per sec. The velocity of the gas stream is 6400 fps. Neglect air resistance and the decrease of rocket weight.
   a. What maximum velocity will the rocket attain?
   b. To what height will it rise?
   c. With what velocity will it strike the earth in falling?

1.4 A single flat surface (vertical plate) is carried at a speed of 10 fps toward a nozzle that discharges water at 8 cfm at 50 fps. What force is required to carry the plate if it is maintained directly toward the nozzle? How much water hits the plate each minute?

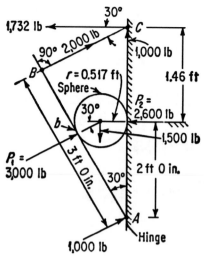

Exhibit 1.2

1.5 A tank truck full of fresh water is proceeding along the road at 40 mph. The tank may be considered to be a cylinder with a diameter of 6 ft and a length of 20 ft. If the truck is suddenly stopped, what will be the longitudinal force in pounds per inch of circumference in the front end of the cylinder wall? What will be the lateral force per inch of length in the tank wall? Neglect static pressure due to water head.

1.6 The position of a body of 70-lb mass is given by the equation $x = 3t^2 + 2t + 4$, where x is in feet and t is in seconds. (a) Compute the velocity when x is equal to 70; (b) What force is required for the motion to take place?

1.7 In the hand-operated press shown in Exhibit 1.7, a force F produces a compressive reaction on the body B placed between the movable platform and the fixed anvil. If $F = 20$ lb, $a = 1$ in., $b = 8$ in., $c = 10$ in., $d = 2$ in., $e = 20$ in., $m = 2$ in., $n = 15$ in., and $x = 5$ in., what will be the compressive reaction on body B?

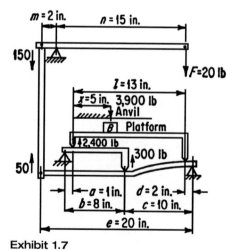

Exhibit 1.7

1.8 A bomber is flying with a constant velocity v at an elevation H toward its objective, which is being sighted by an observer in the plane. At what angle of sight with respect to the vertical should a bomb be released so that it would strike the objective? Neglect air resistance.

1.9   An automobile weighing 2800 lb is traveling at 30 mph when it hits a depression in the road which has a radius of curvature of 50 ft. What is the total force to which the springs are subjected?

1.10   A rocket missile of weight W is to leave the earth's surface and be projected out into space to infinite height. Determine the escape velocity, i.e., the initial velocity required to accomplish this feat.

1.11   A ship is propelled by a constant thrust $p \times m$, where m is the mass of the ship. The water exerts a resistance $c \times m \times v^2$, where p and c are constants.
   a.   What is the terminal velocity?
   b.   Starting from rest, what is the distance traveled? Express the solution as a function of time.
   c.   Repeat part (b) and express the solution as a function of velocity.

1.12   A curved section of railroad track has a radius of 2500 ft. As shown in Exhibit 1.12, track gage is 4 ft $8\frac{1}{2}$ in. and the superelevation of the outside rail is 4 in. What is the maximum speed in miles per hour at which a train may negotiate the curve without having wheel flanges exert side thrust on the rails?

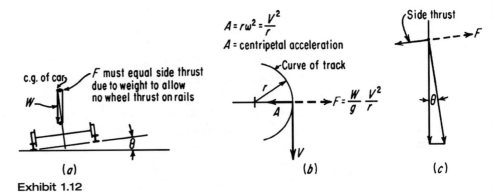

Exhibit 1.12

1.13   A hanger for an equipment platform is to carry a load of 175 kips. Design an eyebar of A-440 steel.

1.14   A rigid horizontal slab of uniform construction weighing 120,000 lb is first centrally supported by a 12-in.-high steel column (Exhibit 1.14). Two aluminum columns 11.90 in. high are symmetrically placed, one on each side of the steel column. All three columns are subjected to a rise in temperature to pick up the entire 120,000-lb load.

Determine to what minimum rise in temperature the three columns must be subjected to pick up the *entire 120,000-lb* load.

Properties of columns:

Steel—area = 10 in.^{2}; $E_s = 30 \times 10^6$; coefficient of expansion 0.0000065 in./(in.) (°F).

Aluminum—area = 20 in.^{2}; $E_a = 9.6 \times 10^6$; coefficient of expansion 0.0000240 in./(in.) (°F).

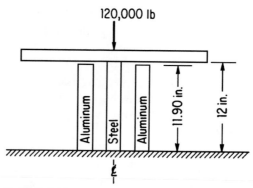

Exhibit 1.14

SOLUTIONS

1.1

$$t = \frac{1}{2}\left(\frac{2v_0}{g}\sin\theta \pm \sqrt{\frac{4v_0^2}{g^2}\sin^2\theta - 4\frac{2hg}{g}} \right)$$

$$= \frac{1}{2}\left(\frac{2\times2000}{32.2}\times0.6 \pm \sqrt{\frac{4\times2000^2}{32.2^2}\times0.36 - 4\times\frac{2\times100}{32.2}} \right)$$

$$= 37.5 \pm 37.4 = 74.9 \text{ sec}$$

The other is imaginary for the problem.

    The first term $(2v_0/g)\sin\theta$ is the time the projectile takes to reach the top of the trajectory. The first term under the square root sign is the same, because it takes the projectile just as long to come down to the same level as it takes to reach the top of the trajectory, if the square root is taken. The second term under the square root sign is a minus because the plateau has shortened the time of flight by that amount. If there were a drop below the initial level and the time of flight were longer as a result, the minus (−) sign would become a plus (+).

1.2

$$Ab = 2'0''$$
$$P_1 \sin 30 = 3000 \times 0.5 = 1500 \text{ lb}$$
$$P_1 = 3000 \text{ lb}$$
$$BC \times 3 = 3000 \times 2 = 6000$$
$$\text{Thus} \quad BC = 2000 \text{ lb}$$

Reaction at A is 1000 lb at 30° with the horizontal.

Note. ∴ AB was given as 3'0''.
Point A to center line of sphere was given as 2'0''.
Angle ABC was given as 90°.
Angle BCA was given as 30°.

1.3   a.   Propelling force $F = W$, rate of gas stream × gas velocity/g.
       This is rate of momentum of gas stream. Therefore,

$$F = 260/32.2 \times 6400 = 52{,}000 \text{ lb}$$

Time during which gas stream acts until fuel is exhausted is

$$t = 6 \times 2000/260 = 46.2 \text{ sec}$$

Resulting force acting on rocket is

$$F_r = F - W_r = 52,000 - 28,000 = 24,000 \text{ lb}$$

Impulse force of rocket during entire interval gas stream acts is same as momentum of rocket at end of impulse. Therefore,

Rocket impulse $\times t$ = weight of rocket $\times$ max. rocket velocity/g
$$24,000 \times 46.2 = 14 \times 2000 \times v_r/32.2$$

Rearrange and solve for v_r, which is found to be equal to 1265 fps. This is maximum rocket velocity upward.

b. Then rise during impulse is

$$s_1 = \frac{v_r - v_o}{2} t = \frac{1265 - 0}{2} \times 46.2 = 29,200 \text{ ft}$$

Rise after impulse is

$$s_2 = \frac{(v_r)^2}{2g} = \frac{1265^2}{64.4} = 25,000 \text{ ft}$$

Total rise is 29,200 + 25,000 = 54,200 ft.

c. Velocity on striking the earth is

$$v_f = (2gs_f)^{1/2} = (64.4 \times 54,200)^{1/2} = 1860 \text{ fps.}$$

1.4
$$2\left[\frac{W}{g}(v_1 - v_2)\right] = 2\left[\frac{8 \times 62.4}{60} \times \frac{1}{32.2} \times (50+10)\right] = 31 \text{ lb}$$

$$8 \times \frac{60}{50} = 9.6 \text{ cfm of water}$$

1.5 Total weight of water in the tank is calculated out to be 35,300 lb. The truck has a speed equal to $\frac{40}{60} \times 88 = 59$ fps. Now, assume it takes 1 sec to effect a dead stop. Then the deceleration is 59 ft per sec^2 and the force exerted by the water on the front end is

$$F = 35,300/32.2 \times 59 = 64,600 \text{ lb.}$$

The longitudinal force in the cylinder wall is given by

$$f_1 = \frac{64,600}{12\pi \times 6} = 286 \text{ lb per in.}$$

The lateral force in the cylinder wall is given by

$$f_2 = 2 \times 286 = 572 \text{ lb per in.}$$

1.6   a.
$$70 = 3t^2 + 2t + 4$$

$$t^2 + \frac{2}{3^t} - 22 = 0$$

$$t = -\frac{1}{3} \pm \sqrt{\frac{1}{9} + 22} = 4.4 \text{ sec}$$

when x is equal to 70 the velocity is $(6 \times 4.4) + 2 = 28.4$ fps. Note that this is obtained from the equation $x = 6t + 2$, derived from the equation given in the problem.

b.   Force required is equal to $M_{\ddot{x}} = (70/32.2) \times 6 = 13$ lb.

1.7   The forces shown in Exhibit 1.7 have been added in accordance with simple calculations involving moments. Thus, in order to find the reaction at point B

$$\frac{20 \text{ lb} \times 15 \text{ in.}}{2 \text{ in.}} = 150 \text{ lb}$$

$$\frac{e \times 150}{c} = \frac{20 \text{ in.} \times 150 \text{ lb}}{10 \text{ in.}} = 300 \text{ lb}$$

$$\frac{b \times 300}{a} = \frac{8 \text{ in.} \times 300 \text{ lb}}{1 \text{ in.}} = 2400 \text{ lb}$$

$$\frac{1 \times 2400}{13 \text{ in.} - 5 \text{ in.}} = \frac{13 \text{ in.} \times 2400}{8 \text{ in.}} = 3900 \text{ lb}$$

1.8   Refer to Exhibit 1.8. Let $L = v_x t$, since velocity is constant. We know it will take just as long for the bomb to fall through H as to cover distance L. Also $H = \frac{1}{2gt^2}$ and by substitution

$$L = v_x \left(\frac{2H}{g}\right)^{1/2}$$

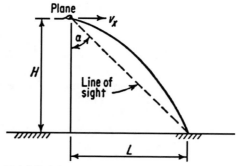

Exhibit 1.8

The angle of sight α may be obtained in terms of tan α so that

$$\tan \alpha = \frac{L}{H} = \frac{v_x}{H} \times \left(\frac{2H}{g}\right)^{1/2}$$

$$\tan \alpha = v_x \times 0.25\left(\frac{1}{H}\right)^{1/2}$$

and

$$\alpha = \tan^{-i} 0.25 \times \frac{v}{H^{1/2}}$$

1.9 The acceleration into the curve is

$$a = \frac{v^2}{\text{radius of curvature}} = \frac{\left(\frac{30}{60} \times 88\right)^2}{50} = 38.7 \text{ ft per sec}^2$$

Let us now equate the vertical forces to obtain the following equation of motion:

$$F - 2800 = \frac{W}{g} 38.7 = \frac{2800}{32.2} 38.7 \quad \text{or} \quad F = 6170 \text{ lb force on the springs.}$$

1.10 The missile must pass out beyond the earth's force of gravity or gravitational pull. We remember that the earth's attraction force F is proportional to the square of the distance x away from the center of the earth to that point of force where it becomes zero. For this purpose it is equal to

$$F = \frac{WR^2}{x^2}$$

where R is the earth's radius. The work required to go from $x = R$ to $x = \alpha$ is by integration of $\int_R^a F \, dx$ is WR. There is a degeneration of initial kinetic energy to zero kinetic energy and this equals the work expended. Initially, the kinetic energy is $\frac{1}{2}(W/g)v_0^2$, and finally it is zero. Then

$$WR = \frac{1}{2}\frac{W}{g}v_0^2$$

where v_0 is the escape velocity in fps. By clearing both sides of the equation and solving for v_0,

$$v_0 = (2gR)^{1/2}$$

1.11 a. Refer to Exhibit 1.11. $\Sigma F = m \times a$. Then from Newton's second law of motion, at terminal velocity, $V = $ constant and therefore $a = 0$.

$$p \times m - c \times m \times v^2 = 0 \text{ and terminal velocity } v = \sqrt{p/c}.$$

Exhibit 1.11

The differential equation of motion is

$$\Sigma F = m \times \frac{d^2 x}{dt^2} = m \times \frac{dV}{dt}$$

$$p \times m - c \times m \times V^2 = m \times \frac{dV}{dt} \quad \text{and} \quad p - c \times V^2 = \frac{dV}{dt}$$

$$\frac{dV}{dt} + c \times V^2 - p = 0$$

Separate variables so that $(dV/p - cV^2) = dt$.

$$\int \frac{dV}{p - cV^2} = \int dt$$

From tables,

$$\frac{1}{2pc} \ln \frac{p + V\sqrt{pc}}{p - V\sqrt{pc}} + C_{\text{constant}} = t$$

When $t = 0$, $V = 0$. Then

$$\frac{1}{2\sqrt{pc}} \ln \frac{p}{p} + c = 0$$

Then since ln (1) = 0, $c = 0$.

$$\ln \frac{p + V\sqrt{pc}}{p - V\sqrt{pc}} = 2\sqrt{pc}\,(t)$$

$$e^2\sqrt{pc}\,(t) = p + V\sqrt{\frac{pc}{p}} - V\sqrt{pc}$$

$$V = \sqrt{\frac{p}{c}} \frac{e\sqrt[2]{pct} - 1}{e\sqrt[2]{pct} + 1}$$

$$V = v \frac{1 - e\sqrt[2]{pct}}{1 + e\sqrt[2]{pct}}$$

where V = terminal velocity.

b. Because $p = dx/dt$,

$$\int dx = \int V \frac{1 - e\sqrt[2]{pct}}{1 + e\sqrt[2]{pct}} \, dt$$

$$\int dx = V \int \frac{1}{1 + e\sqrt[2]{pct}} \, dt - V \int \frac{e\sqrt[2]{pct}}{1 + e\sqrt[2]{pct}} \, dt$$

$$\int dx = V \int \frac{dt}{1 + e\sqrt[2]{pct}} - V \int \frac{dt}{1 + e\sqrt[2]{pct}}$$

$$\frac{x}{V} = \frac{1}{2\sqrt{pc}} \left[2\sqrt{pct} - \ln\left(1 + e\sqrt[2]{pct}\right) \right]$$

$$= -\frac{1}{-2pc} \left[-2\sqrt{pct} - \ln\left(1 + e^{-\sqrt[2]{pct}}\right) \right] + c$$

where c = constant of integration

$$\frac{x}{V} = \frac{1}{2\sqrt{pc}}\left[-\ln\left(1+e^{\frac{2}{3}}\sqrt{pc}t\right) - \ln\left(1+e^{-\frac{2}{3}}\sqrt{pc}t\right)\right] + c$$

When $t = 0$, $x = 0$, and

$$0 = \frac{1}{2\sqrt{pc}}(-\ln e - \ln\ 2) + c$$

Therefore

$$c = \frac{2\ \ln\ 2}{2\sqrt{pc}} = \frac{1.386}{2\sqrt{pc}}$$

Finally

$$x = \frac{V}{2\sqrt{pc}}\left[1.386 - \ln\left(1+e^{\frac{2}{3}}\sqrt{pc}t\right) - \ln\left(1+e^{-\frac{2}{3}}\sqrt{pc}t\right)\right]$$

c. Now going back: $dV/dt + cV^2 - p = 0$. And dividing through by dx/dt,

$$\frac{dV}{dt} + \frac{cV^2}{dx/dt} - \frac{p}{dx/dt} = 0$$

Because $dx/dt = V$,

$$\frac{dV}{dx} + cV - \frac{p}{V} = 0$$

Now separate variables.

$$\int\frac{dV}{-cV + p/V} = \int dx$$

Let $V^2 = \mu$, then $2V\,dV = d\mu$.

$$\int dx = \frac{1}{2}\int\frac{2V\,dV}{p-cV^2} = \frac{1}{2}\int\frac{d\mu}{p-c\mu}$$

$$x = \frac{1}{2}\left[\frac{1}{-c}\ln(p-c\mu)\right] + c$$

And therefore,

$$x = \frac{1}{2}\left[\frac{1}{-c(p-cV^2)}\right] + \text{constant}$$

when $x = 0$, $V = 0$, so that

$$0 = \frac{1}{2}\left[\frac{-1}{c}\ln p\right] + c$$

$$c = \frac{1}{2c}\ln p.$$

Substituting, we have

$$x = \frac{1}{2c}[\ln p - \ln(p - cV^2)].$$

1.12 Refer to Exhibits 1.12b and c. Centripetal force F must equal side thrust due to weight to allow no wheel thrust on rails.
Track gage expressed in inches = 56.5 in.

$$\theta = \arcsin\ 4/56.5 = \arcsin\ 0.0708$$

From vehicle dynamics,

$$\theta = \arctan\ \frac{F}{W} = \arctan\frac{W}{g} \times \frac{V^2/r}{W} = \arctan\frac{V^2}{gr}$$

For small angles tan = sin. Therefore, let $V^2/gr = 4/6.5$ and $V^2 = 5700$ ft²/sec².

$$V = 75.5 \text{ ft per sec or } 51.5 \text{ mph}$$

1.13 Refer to Exhibit 1.13 for the notational system. Let subscripts 1 and 2 refer to cross sections through the body of the bar and through the center of the pinhole, respectively. From the AISC *Manual* for A-440 steel,
If $t \le 0.75$ in. $f_y = 50$ kips/in.².
If $0.75 < t \le 1.5$ in., $f_y = 46$ kips/in.².
If $1.5 < t \le 4$ in., $f_y = 42$ kips/in.².
Design the body of the member, using a trial thickness.

The specification restricts the ratio w/t to a value of 8. Compute the capacity P of a $\frac{3}{4}$-in. eyebar of maximum width. Thus, $w = (8)(\frac{3}{4}) = 6$ in.; $f = 0.6(50) = 30$ kips/in.²; $P = (6)(0.75)(30) = 135$ kips. This is not acceptable because the desired capacity is 175 kips. Hence, the required thickness exceeds the trial value of $\frac{3}{4}$ in. With t greater than $\frac{3}{4}$ in., the allowable stress at 1 is $0.60f_y$, or $0.60(46$ kips/in.²$) = 27.6$ kips/in.², say 27.5 for design purposes. At 2 the allowable stress is $0.45(46) = 20.7$ kips/in.², say 20.5 kips/in.² for design purposes.

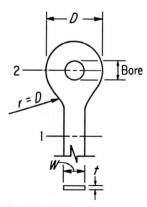

Exhibit 1.13

To determine the required area at 1, use the relation

$$A_1 = \frac{P}{f}$$

where f = allowable stress as computed above. Then

$$A_1 = \frac{175}{27.5} = 6.36 \text{ in.}^2$$

in which $A_1 = 6.5$ in.^{2}. Use a plate $6\frac{1}{2} \times 1$ in.

Design the section through the pinhole. The AISC specification limits the pin diameter to a minimum value of 7 ($w/8$). Select a pin diameter of 6 in. The bore will then be $6\frac{1}{32}$ in. in diameter. The net width required will be $P/ft = 175/(20.5 \times 1.0) = 8.54$ in. And $D_{min} = 6.03 + 8.54 = 14.57$ in. Set $D = 14\frac{3}{4}$ in.; $A_2 = (1.0)(14.75 - 6.03) = 8.72$ in.^{2}. $A_2/A_1 = 1.34$. This result is OK, because the ratio A_2/A_1 must lie between 1.33 and 1.50.

Determine the transition radius r. In accordance with the specification, set $r = D = 14\frac{3}{4}$ in.

1.14 Refer to Exhibit 1.14. Aluminum columns will carry full load when their shortening under load just equals the differential thermal expansion over the steel column. Then, shortening of aluminum columns

$$\Delta_T = \frac{Pl}{AE} = \frac{120 \times 12}{40 \times 9600} = 3.75 \times 10^{-3} \text{ in.}$$

Differential thermal expansion

$$\Delta_T = \Delta \alpha \, \Delta Tl = (24 - 6.5)10^6 \times 12\Delta T = 2.1 \times 10^{-4} \Delta T \text{ in.}$$

Therefore, $\Delta_P = \Delta_T - 0.1$ in.   or $3.75 \times 10^{-3} = 2.1 \times 10^{-4} \Delta T - 0.1$

$$\Delta T = \frac{0.1 + 3.75 \times 10^{-3}}{2.1 \times 10^{-4}} = \frac{1000 + 3.75}{2.1} = 494°F$$

Thermodynamics

OUTLINE

PROBLEMS

2.1 An isentropic convergent nozzle is used to evacuate air from a test cell which is maintained at stagnation pressure of 300 psia and temperature of 800°R. The nozzle has inlet and outlet areas of 2.035 and 1.000 sq ft, respectively; the nozzle is secured to the wall of the test cell with anchor bolts imbedded in the wall. Constant specific heat c_p is to be taken as 0.24; the specific heat ratio is to be taken as 1.4; and the gas constant R is to be taken as 53.35 lb force foot per lb mass °R. Calculate the total tensile load on the anchor bolts.

2.2 Air at a mixture pressure of 20 psia and a temperature of 240°F with a relative humidity of 2 percent is used to heat 5000 lb of solid plastic material prior to a certain manufacturing process. The plastic has a specific heat of 0.36 Btu/(lb)(°F). When placed in the oven, the plastic is at 40°F, and when removed, it is at 100°F. The air flow is controlled in such a way that its leaving temperature is always 100°F.
  a. Calculate the air quantity in pounds required on the assumption that the oven is perfectly insulated from its surroundings.
  b. If the heating process requires 46 minutes, specify the average capacity of the fan in cubic feet per minute.

2.3 Two air streams, *A* and *B*, are mixed in direct contact with each other. These are the pertinent data:

	Streams	
	A	*B*
Mass flow lb dry air per hr	2000	3000
Pressure, psia	5	5
Relative humidity, percent	50	90
Dry-bulb. °F	60	85

Assuming constant pressure conditions, what will be the
a. enthalpy,
b. dry bulb, and
c. relative humidity after mixing?

2.4 Saturated steam at 300 psia flows through a 4-in. standard steel pipe 500 ft long at 1000 lb per hr. The line is insulated with 2 in. of 85 percent magnesia. If the pipe is situated in a 20°F atmosphere, what is the quality of the steam at the end of the 500-ft run?

2.5 A 20,000-kw turbogenerator is supplied with steam at a pressure of 300 psia and a temperature of 650°F. The back pressure is 1 in. Hg abs. At best efficiency the combined steam rate is 10 lb per kwhr.
a. What is the combined thermal efficiency (CTE)?
b. What is the combined engine efficiency (CEE)?
c. What is the ideal steam rate?

2.6 A turbogenerator is operated on the reheating-regenerative cycle with one reheat and one regenerative feed-water heater. Throttle steam at 400 psia and a total steam temperature of 700°F are used. Exhaust at 2 in. Hg abs. steam is taken from the turbine at a pressure of 63 psia for both reheating and feed-water heating. Reheat to 700°F.

For the ideal turbine working under these conditions, find
a. Percentage of throttle steam bled for feed-water heating.
b. Heat converted to work per pound of throttle steam.
c. Heat supplied per pound of throttle steam.
d. Ideal thermal efficiency.
e. And draw the temperature-entropy diagram showing boiler, turbine, condenser, feed-water heater, and piping.

2.7 Determine the force acting on a piston 1 in. in diameter if it is pushed 3 in. into an airtight cylinder 4 ft long. Consider that piston is pushed quickly.

2.8 A compressed-air receiver is maintained at 300 psia and 80°F. Air is led from the receiver through 2-in.-diameter pipe of Schedule 40 thickness. A complete rupture occurs in the line at a distance from the receiver equal to 100 ft equivalent. Assume compressor capacity is sufficient to maintain the above-stated temperature and pressure, and determine the quantity of air which escapes to atmosphere through the ruptured pipe. Assume air as perfect gas with a specific heat ratio of 1.4.

2.9 The vacuum in a surface condenser is 28 in. Hg referred to a 30-in. barometer. The temperature in the condenser is 80°F. Find the percent by weight of air present in this condenser.

2.10 When a pressure vessel contains 1 lb of water vapor at 300°F, the pressure gauge reads 10 psi. If 1 lb of dry air is pumped into the vessel with the temperature of the system remaining the same, what will be the new pressure reading?

2.11 Air discharged from a compressor before cooling to remove moisture is at 150 psig and 150°F. Consider the gas to be saturated with water vapor. Barometer is at 14.0 psia. To what temperature must the gas be cooled at the cooler outlet to prevent condensation in the distribution system where the pressure has dropped to 100 psig and the temperature is 50°F?

2.12 One hundred lb of steam, condensed at 150 psia, is discharged through a trap into the atmosphere without further cooling in the trap discharge line. What percentage of the condensate will "flash" upon discharging? Assume a constant enthalpy expansion and use method of properties of a wet mixture.

2.13 An air compressor compresses air from atmospheric pressure to 70 psig at the rate of 400 cfm. The increase in internal energy is 1200 Btu per min, and 300 Btu per min are rejected during the process through the aftercooler. If the initial specific volume is 13.55 and the final specific volume is 3.53, calculate the work done on the air by the compressor. Hint: Total work is equal to change in internal energy plus heat rejected plus work of compression.

2.14 Ten horsepower are absorbed during the compression strokes of an engine. During 1 min 100 lb of cooling water flowing through the cylinder jacket has its temperature raised 4.2°F by heat absorbed from the cylinder during the compression strokes. (a) Find the change in internal energy of the medium being compressed in Btu per minute. (b) What kind of a process is this?

2.15 Superheated steam is generated at 1350 psia and 950°F. It is to be used in a certain process as saturated steam at 1000 psia. It is desuperheated in a continuous manner by injecting water at 500°F. How many pounds of saturated steam will be produced per pound of original steam?

2.16 Air at 40°F and 2800 psia flows through a $\frac{1}{2}$-in.-ID tube at the rate of 75 scfm. It is expanded through a valve into a $\frac{1}{2}$-in. standard pipe (Sch. 40) to atmospheric pressure. Estimate the temperature of the air a short distance beyond the valve, assuming that a negligible amount of heat is transferred in from the surroundings.

2.17 When nitrogen at 100 atm and 80°F expands adiabatically and continuously to 1 atm through a throttle, the temperature drops to 45°F. A process for liquefying nitrogen involves a continuous flow exchanger system with no moving parts with the gas entering at 100 atm and 80°F, unliquefied gas leaving at 70°F and 1 atm, and liquid drawn off at 1 atm abs. It is estimated that the system is well enough insulated so that the heat leak is reduced to 25 Btu per lb mole of nitrogen entering. It is claimed that 3.5 percent of the entering nitrogen can be liquefied on one pass through the apparatus. Do you think this is correct? Show basis for your answer.

2.18 A 4-ft-diameter duct carrying air of density 0.0736 lb per cu ft is traversed by a pitot tube using the 10-point method. The readings in in. of water at 72°F from one side of the duct to the other are, respectively, 0.201, 0.216, 0.220, 0.219, 0.220, 0.220, 0.218, 0.219, 0.220, and 0.216. Find the average velocity and the mass flow.

2.19   Air at 60°F and atmospheric pressure is compressed, liquefied, and separated by rectification into pure oxygen and pure nitrogen. The two gases are finally compressed into storage cylinders at 2000 psig and 100°F. Calculate change in enthalpy, entropy, and energy for the whole process per 1000 cu ft of air treated. Assume ideal gases. Use units of Btu and degrees F. Assume air to consist only of oxygen and nitrogen.

SOLUTIONS

2.1   See Exhibit 2.1. Because back pressure on the nozzle is considered atmospheric the flow is critical or unretarded. The nozzle reaction force is exerted on the curved portion of the nozzle contour, causing tension in the holding bolts. The critical pressure $P_c = 0.53 \times P_1$. Or $0.53 \times 300 = 159$ psia. On leaving the nozzle the gas expands rapidly and the inertia of the rapidly expanding air stream sets up a series of expansion and compression waves.

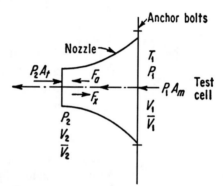

Exhibit 2.1

Now let the following nomenclature apply:

v_1 = gas velocity at mouth of nozzle, fps
v_2 = gas velocity at nozzle throat, fps
G = gas flow rate, lb per sec
P_1 = pressure at nozzle mouth, psia
P_2 = pressure at nozzle throat Pc, psia
A_m = nozzle mouth area, 2.035 sq ft
A_t = nozzle throat area, 1 sq ft
$P_1 A_m$ = force of gas on nozzle mouth, lb
$P_2 A_t$ = force of gas on nozzle throat, lb
F_x = axial component of the force F on the curved portion of the nozzle by the expanding gas; this is equal and opposite to the tension forces on the bolts
F_a = force causing acceleration of the gas, lb

$$F_a = Ma = \frac{G}{g}(v_2 - v_1), \text{ lb}$$

Refer to Exhibit 2.1, and you will note that the force at the throat is against the nozzle throat. Thus

$$F_a = P_1 A_m - P_2 A_t - F_x$$

or $\quad F_x = P_1 A_m - P_2 A_t - F_a = P_1 A_m - P_2 A_t - \dfrac{G}{g}(v_2 - v_1).$

Now we must determine the flow G by

$$G = \frac{0.532 A_t P_1}{\sqrt{T_1}} = \frac{0.532 \times 1 \times 300 \times 144}{\sqrt{800}} = 810 \text{ lb per sec.}$$

In order to calculate the gas velocities at mouth and at throat, we must first find the specific volumes at these two points. Assuming a perfect gas,

$$\bar{v}_1 = \frac{RT_1}{P_1} = \frac{53.35 \times 800}{300 \times 144} = 0.985 \text{ cu ft per lb.}$$

For ideal isentropic expansion of gas,

$$\bar{v}_2 = \bar{v}_1 \left(\frac{P_1}{P_2}\right)^{1/k} = 0.985 \left(\frac{300}{159}\right)^{0.714} = 1.574 \text{ cu ft per lb.}$$

Throat velocity is

$$v_1 = \frac{G\bar{v}_1}{A_m} = \frac{810 \times 0.985}{2.035} = 392 \text{ fps}$$

$$v_2 = \frac{810 \times 1.574}{1} = 1275 \text{ fps.}$$

Finally,

$$F_x = (300 \times 144)2.035 - (159 \times 144)1 - (810/32.2)(1275 - 392)$$
$$= 43,200 \times 2.035 - 22,900 - 22,200 = 42,700 \text{ lb}$$

2.2 The heat absorbed by the plastic is released by the air stream. If water vapor were a perfect gas, its enthalpy (total heat) would depend only on temperature and would be independent of pressure; then, values read from the saturated steam tables could be used without error at pressures other than saturation pressure. Actually, the enthalpy of water vapor varies somewhat with pressure but for most psychrometric calculations the quantity of water vapor is small compared to the quantity of air and a slight error in the enthalpy of water vapor introduces only a negligible error in the enthalpy of air-water-vapor mixtures. Thus, for calculations from 32 to 212°F it is sufficiently satisfactory to take the enthalpy of water vapor directly from saturated steam tables, particularly in view of the fact that the saturation pressure of steam is low over this temperature range.

Above 212°F the saturation pressure of steam increases rapidly and at high temperatures it becomes many times as great as the total pressure— 1 atm—of the air-water-vapor system. Therefore, for temperatures above 212°F, it is preferable to refer to superheated steam tables and to use the enthalpy of air-water-vapor mixtures. The enthalpy of the air-water-vapor mixture $(H_m)_t$ at any temperature t can be calculated from

$$(H_m)_t = (H_a)_t + w(H_{wv})_t \text{ Btu per lb dry air}$$

where

$(H_a)_t$ = enthalpy of dry air at temperature t, Btu per lb
w = humidity, lb water per lb dry air
$(H_{wv})_t$ = enthalpy of water vapor at temperature t, Btu per lb

Proceeding, the heat released by the air stream is the difference in enthalpies between the condition of air in and air out. There is no change in moisture content of the air stream, so that the amount of water vapor in the air originally remains unchanged.
Enthalpy—Air In. First determine the humidity ratio w_i,

$$w_i = \frac{24.97 \times 0.02}{20 - 0.5} \times 0.622 = 0.016 \text{ lb moisture per lb dry air.}$$

From superheated steam tables and dry-air tables in the ASHRAE *Guide* or any other standard handbook

$$(H_m)_t = (0.24 \times 240) + 0.016(1160.4) = 76.1 \text{ Btu per lb dry air.}$$

Here the enthalpy was taken from saturated steam tables because temperature was close to 212°F, acceptable here.
Enthalpy—Air Out. Assume negligible pressure loss of air system. Humidity is the same leaving as entering system.

$$(H_m)_t = (0.24 \times 100) + 0.016(1104.4) = 41.7 \text{ Btu per lb dry air.}$$

Heat released to plastic material is by difference.

$$76.1 - 41.7 = 34.4 \text{ Btu per lb dry air.}$$

a. The weight of air required is

wt plastic × sp ht × (100 − 40) = 5000 × 0.36 × 60 = 108,000 Btu
108,000/34.4 = 3140 lb dry air

b. From molal volume relationship, assume a draw-through fan system with the fan handling air at 100°F. Also neglect the effect of humidity at such low-moisture content.

$$3140 \times \frac{379}{29} \times \frac{460 + 70}{460 + 60} \times \frac{1}{46} = 905 \text{ cfm}$$

Let us see just what difference in air volume will result if we base our calculations on the dry air with the original moisture in it.

$$\frac{24.97 \times 0.02}{20 - 0.5} = 0.0257 \text{ cu ft vapor per cu ft dry air}$$

Because there is no loss or gain of water vapor, this ratio will be constant. The total volume of vapor plus dry air is

$$(905 \times 46 \times 0.0257) + (905 \times 46) = 1070 + 41{,}600$$
$$= 42{,}670 \text{ cu ft } 42{,}670/46$$
$$= 928 \text{ cfm}$$

2.3 Applying the same treatment as in Problem 2.2, we shall first find enthalpies of each stream and then combine them.

a. *For stream A,*

$$\frac{0.25614 \times 0.5}{5 - (0.25614 \times 0.5)} \times 0.622 = 0.0164 \text{ water per lb dry air}$$

Weight of moisture in stream $A = 0.0164 \times 2000 = 32.8$ lb

$(H_m)_t = 14.48 + (0.0164 \times 1087.2) = 32.30$ Btu per lb dry air

Enthalpy of stream $A = 2000 \times 32.30 = 64{,}600$ Btu per hr

For stream B,

$$\frac{0.59588 \times 0.9}{5 - (0.59588 \times 0.9)} \times 0.622 = 0.0746 \text{ lb water per lb dry air}$$

Weight of moisture in air stream $B = 0.0746 \times 3000 = 223.8$ lb

$(H_m)_t = 19.32 + (0.0746 \times 1098.3) = 101.1$ Btu per lb dry air

Enthalpy of stream $B = 3000 \times 101.1 = 304{,}000$ Btu per hr

$$\text{Enthalpy of mixture} = \frac{64{,}600 + 304{,}000}{2000 + 3000} = 73.8 \text{ Btu per lb dry air}$$

b. Dry bulb of the mixture, assuming negligible change in specific heat, is

$$\frac{2000}{5000} \times 60 = 24 \quad \frac{3000}{5000} \times 85 = 51 \quad 24 + 51 = 75 \text{ F.}$$

c. Relative humidity of the mixture at 75°F dry-bulb temperature is

$$\frac{32.3 + 223.8}{5000} = \frac{0.42969x}{5 - (0.42969x)} \times 0.622 = 0.0512$$
$$x = 0.885 \qquad \text{relative humidity} = 88.5 \text{ percent.}$$

2.4 The pressure drop is negligible and we can assume a constant pressure system. Temperature corresponding to 300 psia is 417.33°F. From appropriate tables in the ASHRAE *Guide*, the heat loss through a 4-in. pipe insulated as indicated in the problem will lose 0.32 Btu per hr per linear foot of insulated pipe. Then the heat loss will be for the entire length

$$Q = 0.32 \times 500(417.33 - 20) = 63{,}500 \text{ Btu per hr}$$

With heat of condensation at 300 psia equal to 809.3 Btu per lb condensed, the weight of steam condensed within the line may be determined as

$$63,500/809.3 = 78.5 \text{ lb per hr}$$
$$\text{Percent wet} = (78.5/1000)100 = 7.9. \text{ Then quality is}$$
$$1 - 0.079 = 0.921, \ 92.1 \text{ percent.}$$

This is a reasonable quality for steam lines operating saturated at the source point.

2.5 Refer to Exhibit 2.5.

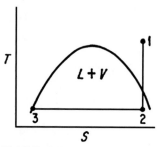

Exhibit 2.5

The combined thermal efficiency is given by

a. $\quad \text{CTE} = \dfrac{3413}{w_r} \dfrac{1}{h_1 - h_3}$

$$= 3413/10 \times 1/(1340.6 - 47.06) = 0.264, \quad \text{or} \quad 26.4 \text{ percent}$$

b. $\quad \text{CEE} = \dfrac{w_i}{w_e} = \dfrac{\text{wt of steam used by ideal engine}}{\text{wt of steam used by actual engine}}$

The weights of steam may also be expressed as Btu per lb. Thus, for the ideal engine the value is 3413 Btu per lb; for the actual engine, $H_1 - H_2$. We already know from the steam tables (or Mollier diagram) the value of H_1. Because the steam expands isentropically into the wet region below the dome of the TS diagram we must first determine the quality at point 2 by calculation. It may also be determined directly by use of the TS diagram for steam. By calculation and application of the method of mixtures

$$S_1 = 1.6508 = S_2 = 0.0914 + x_2 1.9451$$
$$x_2 = \frac{1.6508 - 0.0914}{1.9451} = \frac{1.5594}{1.9451} = 0.80$$
$$H_2 = 47.06 + 0.8 \times 1047.8 = 47.06 + 838 = 885.06 \text{ Btu per lb}$$

$$\text{Ideal steam ram rate } w_i = \frac{3413}{H_i - H_2} = \frac{3413}{1340.6 - 885.06}$$

$$w_i = 3413/455.54 = 7.48 \text{ lb per kwhr}$$
$$\text{CEE} = 7.48/10 \times 100 = 74.8 \text{ percent}$$

2.6 Throttle conditions are designated with subscript 1. From steam tables and the Mollier diagram we have the following:

$$P_1 = 400 \text{ psia}$$
$$t_1 = 700°F$$
$$H_1 = 1362.2 \text{ Btu per lb}$$
$$S_1 = 1.6396$$
$$H_2 = 1178 \text{ Btu per lb}$$
$$H_3 = 1380.1 \text{ Btu per lb}$$

a. Percentage of throttle steam bled for feed-water heating is

$$\frac{\text{Heat added}}{\text{Heat supplied}} = \frac{H_6 - H_5}{H_2 - H_5} = \frac{265.27 - 69.10}{1178 - 69.10} = \frac{196.17}{1107.9}$$

$$= 0.1771, \quad \text{or} \quad 17.17 \text{ percent}$$

b. Heat converted to work per pound of throttle steam is

$$H_1 - H_2 + 0.8229(H_3 - H_4) = 1362.2 - 1178$$
$$+ 0.8229(1380.1 - 1035.8)$$
$$184.3 + 0.8229(344.3) = 184.3 + 283 = 467.3 \text{ Btu per lb}$$

c. Heat supplied per pound of throttle steam is

$$H_1 - H_6 + (H_3 - H_2) = 1362.3 - 265.27 + 1380.1 - 1178$$
$$1097.03 + 202.1 = 1299.13 \text{ Btu per lb}$$

d. Ideal thermal efficiency is

$$\frac{\text{Heat converted to work}}{\text{Heat supplied}} = \frac{467.3}{1299.13} = 0.361, \quad \text{or} \quad 36.1 \text{ percent}$$

e. Refer to Exhibits 2.6a and 2.6b.

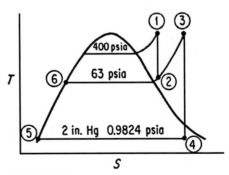

Exhibit 2.6a

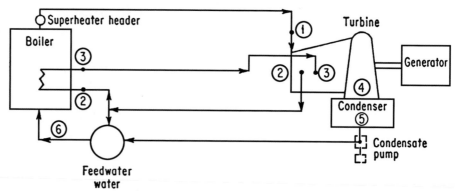

Exhibit 2.6b

2.7   The initial pressure is atmospheric and the initial volume is found to be 37.7 cu in. *before* the piston is moved into the cylinder.

$$0.785 \times 1 \times 1 \times (4 \times 12) = 37.7 \text{ cu in.}$$

After moving 3 in. into the cylinder, the new volume becomes 35.3 cu in.

$$0.785 \times 1 \times 1 \times (3.75 \times 12) = 35.3 \text{ cu in.}$$

Assume the compression is polytropic with the compression exponent n equal to 1.35, then set up the following relation with the final pressure P_2:

$$14.7/P_2 = (0.0204/0.0218)^{1.35} \quad \text{or} \quad P_2 = 16 \text{ psia}$$

Pressure generated within the cylinder is $16 - 14.7 = 1.3$ psi.
  Finally,

$$\text{Force} = 1.3 \times 0.785 \times 1^2 = 1.02 \text{ lb}$$

where $0.0204 = 35.3/1728$ and $0.0218 = 37.7/1728$

2.8   The acoustic velocity sets the maximum flow rate of a gas in a pipe or other conduit. From the acoustic velocity assumed to take place at point of rupture (2-in.-diameter opening) we can find the maximum gas flow rate in cubic feet per second, giving the greatest possible pressure drop in the 100 ft of pipe. If the pressure drop calculated ensures critical flow, then the flow rate becomes fixed. The resulting pressure drop at acoustic velocity when subtracted from the steady receiver pressure will determine the type of flow (retarded or unretarded) into the atmosphere. The acoustic velocity may be determined in accordance with

$$v_e = \sqrt{kgRT} = \sqrt{1.4 \times 32.2 \times 53.5 \times 540} = 1140 \text{ fps}$$

The flow rate at point of rupture is simply and quickly given by

$$\frac{1140 \times 2^2}{183} = 25 \text{ cfs}$$

To ensure critical flow the minimum pressure upstream is

$$P_1 = 14.7/0.53, \quad \text{or} \quad 28 \text{ psia}$$

where 14.7 psia is atmospheric pressure downstream of rupture. For unretarded flow, the flow rate is constant for all conditions of critical flow. To ensure critical flow, the pressure drop through the pipeline must not exceed $300 - 28 = 272$ psi. For unretarded flow of air the flow rate in pounds per second is found to be

$$G = \frac{0.532 A_0 P_1}{\sqrt{T_1}} = \frac{0.532 \times 0.785 \left(\frac{2}{12}\right)^2 \times 28 \times 144}{\sqrt{540}} = 2.1 \text{ lb per sec.}$$

This shows the density of the air stream at point of exit to be 2.1/25, or 0.084 lb per cu ft. The free air flow rate is also found to be equal to $2.1/0.075 = 28$ cfs. This is also equivalent to 28×60, or 1680 cfm. From pressure-drop charts in standard manuals or standards of the Compressed Air Institute, the pressure drop for the problem at hand and for the conditions given is found to be approximately 8 psi per 100 ft of 2-in. pipe. Thus, the flow remains critical and the flow rate is constant at 2.1 lb per sec.

2.9 At this temperature 1 in. Hg exerts a pressure of 0.4875 psi. Also, if the vacuum as stated in the problem were 30 in. Hg referred to a 30-in. barometer, the absolute pressure in the condenser would be zero. In a condenser, steam (water vapor) is condensing in contact with water and may be taken as saturated.

$$\frac{0.5067}{(2 \times 0.4875) - 0.5067} \times 0.622 = 0.672 \text{ lb water per lb dry air}$$

$$\frac{1}{1 + 0.672} \times 100 = 60 \text{ percent by weight of air}$$

2.10 The problem assumes no water vapor condensation. Assume molecular weight of the air is 29, that of water vapor 18. We will work this solution by finding the vessel volume by assuming the application of the perfect-gas law. Add the 1 lb of air to the 1 lb of water vapor, and with the volume and temperature known we can determine the new pressure.

$$p_1 V = NKT = (10 + 14.7)V = \left(\frac{1}{18}\right)(10.71)(460 + 300)$$

Then $$V = \frac{\left(\frac{1}{18}\right)(10.71)(760)}{24.7} = 18.4 \text{ cf vessel volume}$$

The average molecular weight of the mixture of water vapor and dry air is

$$M_m = \frac{2}{\left(\frac{1}{18}\right) + \left(\frac{1}{29}\right)} = 22.2.$$

Set up the perfect-gas law formula to solve for the new pressure, p_2.

$$p_2 \times 18.4 = (2/22.2)(10.71)(760)$$
$$p_2 = (2/22.2)(10.71)(760)/18.4 = 40 \text{ psia,} \quad \text{or} \quad 25.3 \text{ psi}$$

2.11 From steam tables, at 150°F the saturated vapor pressure is 3.716 psia. At the compressor discharge before cooling, the vapor content is determined

$$3.716/(150 + 14 - 3.716) = 0.0232 \text{ cu ft vapor per cu ft dry air.}$$

At a remote point where the temperature will be 50°F, the saturated vapor pressure is 0.178 psia, and the vapor content is

0.178/(100 + 14 − 0.178) = 0.00156 cu ft vapor per cu ft dry air.

This last figure is the maximum amount of water vapor content at the remote point and this must be the vapor content at the cooler discharge to prevent condensation. The cooler outlet temperature can now be calculated by determining the saturated vapor pressure at the remote point:

$$\frac{p}{150+14-p} = 0.00156$$

from which p is found to be 0.2554 psia. Refer to steam tables, saturated steam section. For a pressure of 0.2554 psia the corresponding saturated temperature is approximately 60°F. This is the cooler outlet temperature desired.

2.12 Use saturated steam tables and check with Mollier diagram. Enthalpy of initial condensate before expansion is equal to the sum of the enthalpies (saturated vapor and saturated liquid) in the final wet mixture.

$$H_1 = H_f + XH_{fg} = 330.53 = 180.07 + X970.3$$
$$X = \frac{330.53-180.07}{970.3} = 0.015 \quad \text{or} \quad 1.5 \text{ percent}$$

2.13 The work done is

$$A(P_1V_1 - P_2V_2) = \frac{1}{778} \times 144[(14 \times 400) - \{(70 + 14 \times 400 \times 3.53/13.55)\}]$$
Work done = 1200 + 300 + 600 = 2100 Btu

2.14 a.   10 × 33,000/778 = 424 Btu per min

Heat absorbed by 100 lb water: $100 \times 1 \times 4.2 = 420$ Btu.

This is equivalent to 420 × 778 = 326,900 ft-lb. Change in internal energy is equal to heat removed minus heat of compression, that is, 420 − 424 = −4 Btu per min.

b.   This is a polytropic process for gases.

2.15 Enthalpy steam at 1350 psia and 950°F (H_1) = 1465 Btu per lb. Enthalpy saturated steam at 1000 psia (H_2) = 1191 Btu per lb. Enthalpy water at 500°F (H_3) = 488 Btu per lb. Now let X equal lb of 500°F water required. Then

$$H_1 + XH_3 = (1+X)H_2$$
$$X = \frac{H_1-H_2}{H_2-H_3} = \frac{1465-1191}{1191-488} = 0.39$$

Thus, 1 + 0.39 = 1.39 lb saturated steam produced per lb original steam.

2.16 Assume a constant enthalpy process. Then

$$H_1 - \frac{v_1^2}{2g\left(\frac{1}{778}\right)} = H_2 - \frac{v_2^2}{2g\left(\frac{1}{778}\right)}.$$

Cfm at condition 1 is $75 \times \left(\dfrac{15}{2800}\right) \times \left(\dfrac{500}{520}\right) = 0.385$

$$v_1 = \frac{0.385 \times 144}{60 \times 0.25 \times 0.785} = 4.7 \text{ fps}$$

Now estimate t_2 equal to $-50°$F. Then

$$v_2 = \left(\frac{75}{60}\right) \times (144/0.304) \times \left(\frac{410}{520}\right) = 470 \text{ fps}$$

$$H_2 = H_1 - \left(\frac{1}{778} \times 1/64.4 \times 470^2\right) = 102 - 4.4 = 97.6 \text{ Btu per lb}$$

Temperature from enthalpy chart at 97.6 is $-48°$F.

2.17 Datum is at $-460°$F vaporization where enthalpy is zero. Enthalpy (H_1) of nitrogen gas at 100 atm and 80°F = 125 Btu per lb. Enthalpy (H_2) of nitrogen gas at 1 atm and 70°F = 130 Btu per lb.

$$Q = \frac{25}{28} = \text{heat leak Btu per lb of entering nitrogen}$$

Enthalpy (H_3) of liquid nitrogen at 1 atm and $-322°$F (77.4 Kelvin)

$$H_3 = 0.24(460 - 322) - (1335 \times 1.8/28) = -53 \text{ Btu per lb}$$
$$Q + H_1 = (1 - X)H_2 + XH_3$$
$$X = \frac{H_2 - H_1 - Q}{H_2 - H_3} = \frac{4.11}{183} = 0.0224, \quad \text{or} \quad 2.24 \text{ percent liquefied.}$$

Only 2.24 percent of the entering nitrogen can be liquefied. The claim of 3.5 percent is incorrect.

2.18 Average the square root of the readings:

$$\frac{\sqrt{0.210} + \sqrt{0.216} + \cdots + \sqrt{0.216}}{10} = 0.4668 = \sqrt{z_m}$$

The densities of water and air are 62.3 lb pr cu ft and 0.0736 lb per cu ft respectively. The mean $z_m = 0.216$ in. H_2O. This represents a head in ft of fluid flowing as

$$H = \frac{z_m(p_m - p_1)}{p_1} = \frac{0.216(62.3 - 0.074)}{12(0.074)} = 15.1 \text{ ft mass flow.}$$

Finally, the average velocity is

$$V = \sqrt{2gH} = \sqrt{64.4(15.1)} = 31.2 \text{ fps.}$$

2.19 Change in enthalpy is $H_{air} - H_{02} - H_{n2}$. Thus

$$\Delta H = 125 - \left(\frac{32}{39} \times 0.21 \times 109 \right) - \left(\frac{28}{29} \times 0.79 \times 128 \right)$$

$$125 - 25.4 - 98.6 = \text{Approximately zero}$$

Entropy for air $(0.79 \times 0.95 \times 28) + (21 \times 0.88 \times 32)$
    $- 1.987(0.79 \ln 0.79 + 0.21 \ln 0.21)$
    $= 28$ Btu/(lb mol)(°F)
Entropy for nitrogen $0.79 \times 0.605 \times 28 = 13.4$
Entropy for oxygen $0.21 \times 0.58 \times 32 = 3.9$
Mole air $= 1000/379 = 2.64$

Change in entropy for above $= -2.64(28 - 17.3) = -28.6$ Btu/1000 cu ft air
    Energy change $\Delta H - \Delta PV = -\Delta PV = 2.64R(T_1 - T_2)$
$$= 2.64 \times 1.987 \times (60 - 100) = -210 \text{ Btu}$$

CHAPTER 3

Fluid Mechanics

OUTLINE

PROBLEMS 27

SOLUTIONS 31

PROBLEMS

3.1 An elevated water tank consists of a cylindrical section 16 ft in diameter and 20 ft high (Exhibit 3.1). Below this is a hemispherical bottom. With water level 2 ft below the top, a 12-in.-diameter circular orifice with discharge coefficient of 0.60 is opened in the bottom at the center. Compute the time required to draw the water level down 22 ft.

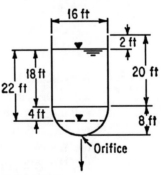

Exhibit 3.1

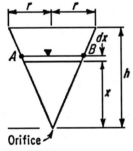

Exhibit 3.2

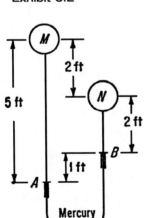

Exhibit 3.3

3.2 A vessel in the form of a right circular cone is filled with water (Exhibit 3.2). If h is its height and r the radius of its base, what time will it require to empty itself through an orifice of area a at the vertex?

3.3 Two sources of pressure, M and N, are connected by a water-mercury differential gauge, as shown in Exhibit 3.3. What is the difference in pressure between M and N in psi?

3.4 It is necessary to pump 3 cfs of water through a pipeline 4000 ft long to a reservoir 200 ft above the pumps. What minimum size of cast-iron pipe (new) should be installed if the loss head is not to exceed 10 percent of the static lift?

27

3.5   Water enters a turbine through a 4-in.-diameter pipe at 150 psia and leaves through an 8-in.-diameter pipe at a point 3 ft lower under a pressure of 5 psig (Exhibit 3.5). Flow is 2 cfs. Turbine efficiency is 85 percent. Compute horsepower output of turbine.

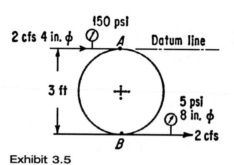

Exhibit 3.5

3.6   The pipe in Exhibit 3.6 is 30 in. in diameter and discharges water at 20 cfs. At point A in the pipe the pressure is 30.5 psig and the elevation is 100 ft. At point B in the pipe, which is 5000 ft along the pipe from point A, the pressure is 50.2 psia and the elevation is 80 ft. Compute the value of the friction factor.

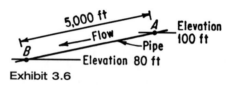

Exhibit 3.6

3.7   A hydraulic test on a needle valve having a $\frac{7}{32}$-in.-diameter seat showed that a flow of 1.485 gpm of water at 60°F caused a drop in pressure through the valve of 2.4 psi. It is intended that this valve be used for compressed-air service and the drop anticipated was to be about 5 psi when connected to the air. What would be the pressure drop in psi if the anticipated air flow is 2710 scfh (14.7 psia and 60°F) when at an internal pressure of 325 psig and 70°F upstream of the needle valve? Is the chosen valve acceptable?

3.8   Steam at an average pressure of 2000 psia and a temperature of 1000°F flows in a pipeline whose inside diameter (ID) is 16 in. If the flow rate is 1,500,000 lb per hr, calculate the pressure drop per 100 ft of pipe. Note that many of the simplified formulas are not applicable at an elevated temperature and pressure. The solution must take into account various friction parameters such as viscosity and Reynolds number.

3.9   A pipe carrying oil having a specific gravity of 0.875 changes in size from 6 in. internal diameter at section A to 15 in. internal diameter at section B (Exhibit 3.9). Section A is 10 ft lower than section B. The pressure in the pipe is 12.00 psi at A and 8.50 psi at B. The discharge is 4.50 cfs. Determine whether the flow is upward or downward.

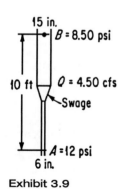

Exhibit 3.9

3.10 The oil pipe in Exhibit 3.10 has a horizontal 60° bend which also reduces the pipe size from 18 to 12 in. What is the magnitude and direction of the total resultant force on the bend when 20 cfs of oil, having a specific gravity of 0.85, is entering the bend per second at a pressure of 8 psi?

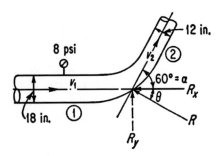

Exhibit 3.10

3.11 At what rate is the surface of the water rising in a vessel whose form is that of an inverted right circular cone when the water is 20 ft deep and is flowing in at a uniform rate of 40 cfm?

3.12 Water is being pumped up a 5 percent grade through a 12-in. cast-iron pipe at 5 cfs. If a pressure gauge at a certain point in the pipe reads 50 psi, what will the pressure reading be 200 pipeline ft farther upstream?

3.13 Exhibit 3.13 shows an oil-speed indicator. The relation between speed and head h is given by: rpm $= C \times h^n$. Determine the calibration constants C and n.

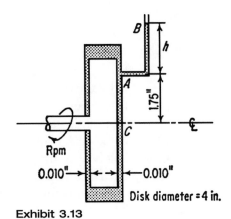

Exhibit 3.13

3.14 Pressurized vessels are often used in plant practice to control the injection rate of a liquid into a process system. When connected to a source of compressed gas, a pressure vessel with an orifice for liquid discharge will eject liquid at a controlled rate. The rate at which the liquid is forced through the orifice depends on the orifice diameter, gas pressure, and liquid density.

Kerosene is to be injected into a pressurizing system by means of nitrogen gas under a constant pressure of injection. The injection takes place through an orifice. Determine the flow rate of injection.

Data and Assumptions:

Sp gr kerosene = 0.8.
Pressure of processing system = 14.7 lb/in.^{2} abs.
Injection pressure constant at 100 lb/in.^{2} gauge.
Orifice diameter = 0.5 in.
Energy losses in the vessel, lines, and orifice are negligible.
Expanding gas behaves as an ideal gas in isothermal conditions, i.e., pressure × volume = constant, when its reduced pressure is relatively low.

Liquid is incompressible and contains no dissolved gases

Liquid flowing from a pressurized vessel enters the process system at ambient pressure, i.e., 14.7 lb/in.^{2} abs.

Pressures of the gas and liquid in a container are identical.

Basic equation

$$Q = 29.85\, Kd_0 \left| \frac{\sqrt{P}}{\sqrt{\text{sp gr}}} \right|$$

where

Q = flow rate of liquid, gal/min
K = orifice discharge coefficient
d_0 = diameter of orifice, in.
P = pressure in vessel (constant), lb/in.^{2} gauge

3.15 Water is being pumped from a deep well through an open-ended pipeline to develop the aquifer. Pipeline length from the well pumps to the plant site is 1000 ft, and the flow velocity in the pipeline is 5 ft/s at a pressure of 250 lb/in.^{2} gauge. Water temperature is 60°F. During the test a plant attendant shut off the flow of water by turning the handle on a plug valve at the end of the open-ended pipeline instantaneously, and as a result the pipe connection back at the pump discharge was ruptured.
a. Determine the pressure rise and the peak line pressure developed in the pipeline by the sudden shutoff.
b. If the valve had been closed gradually, say in 5 s, what would have been the pressure rise and peak line pressure under those conditions? In both cases, neglect the elasticity of water and the pipe.

3.16 A siphon piping system is connected to a reservoir as shown in Exhibit 3.16. Determine the maximum height in feet that can be used for the siphon in a water system if the length of the pipe from the water source to its highest point is 500 ft, the water velocity is 13 ft/s, pipe is 10 in. in diameter, the water temperature is 70°F, and the flow is 3200 gal/min.

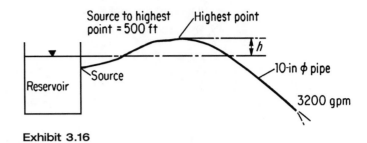

Exhibit 3.16

3.17 Determine the minimum wall thickness and schedule number for a branch steam pipe operating at 900°F if the internal steam pressure is 1000 lb/in.^{2} gauge. Use ANSA B31.1 Code for Pressure Piping and the ASME Boiler and Pressure Vessel Code values and equations where they apply. Steam flow rate is 72,000 lb/h.

3.18 A storage tank containing a flammable liquid is to be provided with a quick-opening valve to an underground sump for draining in the event of a fire. The tank is 20 ft in diameter, and a 20-ft depth of liquid is normally stored. The valve and dumping system is assumed to be equivalent to a 6-in.-diameter orifice. The liquid has a density of 50 lb/ft^3. Also assume a constant orifice coefficient. How long will it take to empty the tank?

3.19 An elevated tank supplies water to a closed process vessel, the flow being due to gravity alone. It is desired to double the rate of flow by increasing the size of the pipe used. One engineer states this can be accomplished by doubling the size of the pipe in diameter. Another claims that the pipe cross section should be doubled. You are asked to settle this dispute. Present your decision, together with supporting calculations. You may assume a constant water level in the tank and in the process vessel, turbulent flow in the pipe, negligible kinetic energy, and negligible entrance and exit losses. Estimate the Fanning friction factor from the equation: $f = (0.0460)/(N_R \times 0.2)$.

3.20 A 3-in. sharp-edged orifice is located in a 6-in.-diameter pipe. A manometer connected to the orifice taps reads 1.2 in. Hg. The pipe is transporting water at 70°F through 500 ft of pipe to the top of a building. The total static head on the pump is 50 ft. Determine the motor horsepower required, assuming an overall efficiency of 78 percent.

SOLUTIONS

3.1 The total time to lower the level may be broken up into two sections: in the cylindrical section and in the hemispherical section. The time for the lowering in the cylindrical section is

$$t_1 = \frac{2A}{CA_0\sqrt{2g}}\left(\sqrt{h_1} - \sqrt{h_2}\right)$$

$$= \frac{2 \times 0.785 \times 16^2}{0.60 \times 0.785 \times \sqrt{64.4}}\left(\sqrt{26} - \sqrt{8}\right) = 242 \text{ sec}$$

The time to lower water in a hemispherical container is given by the general formula

$$t_2 = \frac{\pi}{CA_0\sqrt{2g}} \left[\frac{4}{3}Rh^{3/2} - 0.4h^{5/2} \right]_{h_2}^{h_1}$$

where R is the radius of the hemispherical container and h_1 and h_2 are the respective levels. Then for the problem at hand

$$t_2 = \frac{\pi}{CA_0\sqrt{2g}} \left[\frac{4}{3} \times 8(8^{3/2} - 4^{3/2}) - 0.4(8^{5/2} - 4^{5/2}) \right]$$

$$= \frac{\pi}{0.60 \times 0.785 \times 64.4}$$
$$[10.67(22.6 - 8) - 0.4(181 - 32)]$$
$$= 0.819(156 - 59.5) = 0.819 \times 96.5 = 79 \text{ sec.}$$

Total time $t_1 + t_2 = 242 + 79 = 321$ sec,   or   5.4 min

3.2   Velocity through orifice is

$$v_0 = \sqrt{2gx}$$

Denote by dQ the volume of water discharged in time dt, and by dx the corresponding fall of surface. The flow through the orifice is

$$Q = a\sqrt{2gx} \text{ cfs}$$

being measured as a right cylinder of area of base a and altitude. Therefore, in time increment dt

$$dQ = a\sqrt{2gx}\, dt. \tag{3.1}$$

Denoting S as the area of the surface of the water when depth is x, we have, from geometry,

$$\frac{S}{\pi r^2} = \frac{x^2}{h^2} \quad \text{or} \quad S = \frac{\pi r^2 x^2}{h^2}.$$

But the volume of water discharged in time dt may also be considered as the volume of a cylinder AB of area of base S and altitude dx, hence

$$dQ = S\, dx = \frac{\pi r^2 x^2 dx}{h^2} \tag{3.2}$$

Equating Equations 3.1 and 3.2 and solving for dt,

$$dt = \frac{\pi r^2 x^2 dx}{ah^2\sqrt{2gx}}$$

$$t = \frac{\pi r^2 x^2 dx}{ah^2\sqrt{2gx}} = \frac{2\pi r^2 \sqrt{h}}{5a\sqrt{2g}}$$

3.3 P_m + pressure equivalent to head over A = P_n + pressure equivalent to head over B + pressure equivalent of mercury head (AB). Then $P_m + (5 \times 62.4) = P_n + (2 \times 62.4) + (1 \times 62.4 \times 13.6)$

$$P_m - P_n = 662.8 \text{ psf}, \quad \text{or} \quad 662.8/144 = 4.6 \text{ psi}$$

3.4 The following relation and equation are involved:

$$\text{Velocity} = \frac{Q}{A} = \frac{Q}{0.785d^2} = \text{fps} = \frac{Q}{\pi d^2/4}$$

$$h_f = f_1 \frac{L}{d} \frac{v^2}{2g} = f_1 \frac{L}{d}\left(\frac{4Q}{4d^2\pi}\right)^2 \times \frac{1}{2g}$$

Then

$$d^5 = \frac{8f_1 LQ^2}{\pi^2 g h_f} = \frac{8 \times 0.02 \times 4000 \times 3^2}{\pi^2 \times 32.2 \times (200 \times 0.10)} = 0.907 \text{ ft}$$

$$d = 11.77$$

Use 12-in. cast-iron pipe.

3.5 The flow velocity through the 4-in. pipe may be found to be 23 fps; that through the 8-in. exit pipe, 5.75 fps. Now apply Bernoulli's equation.

$$Z_A + \frac{v_A^2}{2g} + \frac{P_A}{w_A} + h_f + \text{TDH} = Z_B + \frac{v_B^2}{2g} + \frac{P_B}{w_B}$$

$$0 + (23)^2/64.4 + (150 \times 144)/62.4 + 0 + \text{THD} = -3$$

$$+ (5.75)^2/64.4 + (5 \times 144)/62.4$$

$$0 + 8.24 + 346 + \text{THD} = -3 + 0.513 + 11.55$$

$$\text{TDH} = 8.24 + 346 + 3 - 0.513 - 11.55 = 345.18 \text{ ft}$$

$$\text{hp output} = \frac{2 \times 60 \times 62.4 \times 345.18}{33,000} \times 0.85 = 66.5$$

The static head of 3 ft was taken as negative because datum was passed through the supply line. The horsepower equation above is in the form

$$\text{Horsepower} = \frac{\text{lb per min} \times \text{total dynamic head (TDH)}}{33{,}000}$$

3.6 Again, this is a "natural" for Bernoulli's equation. The equation may be modified to suit as follows:

$$h_f = \frac{P_A - P_B}{w} + (Z_A - Z_B) + \frac{v_A^2 - v_B^2}{2g}.$$

Because flow areas are the same with constant pipe size,

$$v_A = v_B = \frac{Q}{(30^2 \times 0.785)/144} = \frac{20}{4.9} = 4.08 \text{ fps}$$
$$P_A = (30.5 \times 144) = 4390 \text{ psf} \quad P_B = (50.2 - 14.7)144$$
$$= 5110 \text{ psf}.$$

Therefore

$$h_f = \frac{4390 - 5110}{62.4} + (100 - 80) = -11.5 + 20 = 8.5 \text{ ft}.$$

From the Darcy equation

$$f_1 = \frac{h_f}{L/D \times v_A^2/2g} = \frac{h_f \times D \times 2g}{L \times v_A^2} = \frac{8.5 \times \left(\frac{30}{12}\right) \times 64.4}{5000 \times 4.08^2} = 0.0164.$$

3.7 Use equation $Q = CA\sqrt{2g\,\Delta h}$. The discharge coefficient C applies to all fluids for all practical intents and purposes. For the water,

$$\frac{1.485}{60 \times 7.5} = C \frac{\left(\frac{7}{32}\right)^2 \times 0.785}{12^2} \sqrt{64.4 \times 2.4 \times 2.31}.$$

Solve for C. It is found to be equal to 0.677, a realistic figure. This value of C is also good for the air flow here. However, flow formula for air must be checked for type of flow, retarded or unretarded.

$$G = (2710/379) \times (29/3600) = 207/3600 = 0.58 \text{ lb per sec}$$
$$P_c = 0.53 \times P_1 \quad P_1 = 325 + 14.7 = 339.7 \text{ psia}$$
$$P_2 = 339.7 - 5 \text{ (assumed)} = 334.7 \text{ psia}$$

Note: The specification of a 5-psi drop through the valve was given by the valve manufacturer to select the valve. The entire purpose of this solution is to check the water-flow and pressure-drop figures submitted by the manufacturer.

$$P_c = 0.53 \times 339.7 = 180 \text{ psia} \quad \text{thus, } P_2 > P_c, \text{ retarded flow}$$

Using the equation of flow that applies for retarded flow, the flow rate at actual operating conditions is found to be

$$G = \dfrac{\dfrac{2.056}{\sqrt{460+70}} \times \dfrac{0.785 \times 0.218^2}{144}}{\sqrt{\left(\dfrac{334.7}{339.7}\right)^{1.43} - \left(\dfrac{334.7}{339.7}\right)^{1.71}}} \times 0.677 = 0.054.$$

This flow rate of 0.054 lb per sec is based on a 5 psi drop. Because the actual flow rate is equivalent to 0.058 lb per sec and flow is considered to be turbulent, the anticipated drop is found as follows applying the laws of affinity:

$$5/x = (0.054/0.058)^2 = 0.885 \quad x = 5/0.885 = 5.67 \text{ psi.}$$

Valve is acceptable.

3.8 First step in any problem involving friction loss is to determine the Reynolds number.

$$Re = \frac{Dv\rho}{\mu} = \frac{\left(\frac{16}{12}\right) \times 118 \times 2.54}{0.03 \times 0.000672} = 19.7 \times 10^6$$

where

$$\text{Velocity of steam} = \frac{Q}{A} = 164/1.39 = \frac{(0.394 \times 1,500,000)/3600}{(\pi \times 1.33^2)/4} = 118 \text{ fps}$$

$$\text{Density of steam} = 1/\text{sp vol} = 1/0.394 = 2.54 \text{ lb per cu ft}$$

Now we see that flow is definitely turbulent. From the Fanning correlation the friction factor is found to be 0.0035. Then the pressure drop is

$$\Delta P/100 \text{ ft} = \frac{2 \times 0.0035 \times 100 \times 2.54 \times 118^2}{144 \times 32.2 \times \left(\frac{16}{12}\right)} = 4 \text{ psi.}$$

The viscosity of steam was found in Marks, *Mechanical Engineers' Handbook* (6th ed., pp. 4–66). Specific volume of steam was obtained from the steam tables.

3.9 Head at

$$B = 8.5 \times 2.31/0.875 = 22.4 \text{ ft}$$

The velocity head at B is calculated as 0.2 ft. Then the total energy head at B is, if datum is taken through B,

$$0 + 0.2 + 22.4 = 22.6 \text{ ft.}$$

Head equivalent to 12 psi is 12 × 2.31/0.875, or 31.8 ft. The velocity head at the same point may be calculated as 8.25 ft. Then the total energy at A is

$$-10 + 8.25 + 31.8 = 30.05 \text{ ft.}$$

Thus we see that the head at A is greater than at B. Therefore, flow is upward.

3.10 By standard calculations for 20 cfs the velocity v_1 in the 18-in. line is 11.31 fps; that in the 12-in. line, 25.5 fps as v_2. Velocity heads: 18-in. pipe $(11.31)^2/2g = 1.99$ ft oil; 12-in. pipe $(25.5)^2/2g = 10.1$ ft oil. Pressure head at point 1 is

$$(8/62.4 \times 0.85)144 = 21.7 \text{ ft oil.}$$

From Bernoulli's equation

$$\frac{P_2}{w} = \frac{P_1}{w} + \frac{v_1^2}{2g} - \frac{v_2^2}{2g}$$

$$21.7 + 1.99 - 10.1 = 13.6 \text{ ft oil.}$$

This is so because friction head is negligible and potential heads equate each other because the bend is in the horizontal plane. The pressure in the 12-in. end of the bend is P_2, which we shall now determine.

$$P_2 = 13.6 \times 62.4 \times 0.85 = 722 \text{ psf}$$

The axial component $R_x = \dfrac{Qw}{g}(v_1 - v_2 \cos\alpha) + (P_1 A_1 - P_2 A_2 \cos\alpha)$

$$R_x = (20 \times 62.4 \times 0.85/32.2)(11.31 - 25.5 \times 0.5) + (1152 \times 0.785 \times 18^2/12^2)$$
$$- (722 \times 0.785 \times 0.5) = 1710 \text{ lb}$$

$$R_y = \left(\frac{Qwv_2}{g} + P_2 A_2\right)\sin\alpha[(20 \times 62.4 \times 0.85 \times 25.5/32.2)$$
$$+ (722 \times 0.785 \times 1^2)]0.866 = 1217 \text{ lb}$$

$$R = \sqrt{R_x^2 + R_y^2} = \sqrt{1710^2 + 1217^2} = 2100 \text{ lb}$$

Direction in which R operates: $\theta = \arctan(R_y/R_x) = 35°25'$

3.11 Volume of vessel:$1/3 \times H \times \pi H^2 = \pi H^3/3$. Then rate of change of volume with respect to height is $dV/dH = (3\pi H^2)/3 = \pi H^2$. Given that $dv/dH = 40$, now $dV/dt = dV/dH \times dH/dt$. Or, $40 = \pi H^2 \times dH/dt$. When H is 20 ft, $dH/dt = (40)/(\pi \times 400) = 0.03183$ fpm, the rate of rising of the water level.

3.12 Vertical lift is (5/100)(200), or 10 ft. Velocity is 6.35 fps calculated. Darcy friction factor f_1 may be taken as 0.02. Then,

$$h_f = 0.02 \times 200 \times 6.35^2/64.4 = 2.51 \text{ ft head loss.}$$

From Bernoulli's equation, $P_A/w_A = 10 + (P_B/w_B) + 2.51 = 115.5$.

$$\frac{Pw}{w_B} = 115.5 - 10 - 2.51 = 103 \text{ ft of water}$$

Finally, $103/2.31 = 44.5$ psi.

3.13 Write the energy equation at points A and B in Exhibit 3.13. The datum plane is at point C.

$$\frac{1.75}{12} + h + 0 = \frac{1.75}{12} - \frac{v^2}{2g} + \frac{P_A}{w}$$

Because P_A/w is small, it may be neglected. Then

$$h = \frac{v^2}{2g} \quad \text{or} \quad v = \sqrt{2gh}$$

$$v = 2\pi r \times \text{rpm}/60 \quad \text{with} \quad r = 1.75/12$$

$$\text{Hence rpm} = [(60 \times 12)/(2\pi \times 1.75)] \times \sqrt{64.4} \times \sqrt{h}$$

$$\text{rpm} = 65.5 \times 8.03\sqrt{h} = 526\sqrt{h}$$

Or, because $\text{rpm} = C \times h^n$, $C = 526$ and $n = 1/2$.

3.14

$$Q = 29.85 \times 0.6 \times 0.5 \frac{\sqrt{100}}{\sqrt{0.8}} = 100 \text{ gal/min}.$$

Usually, a specific flow rate is required for a liquid injection. The above empirical formula may be used with the necessary variables. In a single-charge system, gas pressure diminishes and the liquid flow rate also decreases from the initial discharge rate as the liquid leaves the vessel. Here a different equation is used.

$$Kd_0^2 \frac{t}{v_{g1}\sqrt{\text{sp gr}}} = -0.00228 \, (P_1 + 14.7) \frac{\sqrt{P_2}}{P_2 + 14.7} - \frac{\sqrt{P_1}}{P_1 + 14.7}$$

$$+ 0.261 \tan^{-1} \frac{\sqrt{P_2}}{3.834} - 0.261 \, \tan^{-1} \frac{\sqrt{P_1}}{3.834}$$

where

v_{g1} = initial volume of gas, gal
P_1 = initial pressure of gas, lb/in.^{2} gauge
P_2 = final pressure of gas, lb/in.^{2} gauge

The quantity of liquid injected during a specific time period is found from $V = v_{g2} - v_{g2}$, gal.

3.15 a. Effect of sudden shutoff:

$$\text{Pressure rise} = \frac{\rho c v}{g} = \frac{62.4 \times 4701 \times 5}{32.2 \times 144} = 316 \text{ lb/in.}^2$$

Peak line pressure = 250 + 316 = 566 lb/in.^{2} gauge

b. Gradual closure:

$$\text{Pressure rise} = \frac{\rho L}{g}\left(-\frac{dv}{dt}\right)$$

where

p = water density = 62.4 lb/ft^3
L = length of pipeline = 1000 ft
$-dv/dt$ = deceleration of the liquid stream = 5/5 = 1 ft/s^2

Assume $-dv/dt$ is constant or $-dv/dt = -1$ ft/s^2. Then

$$\text{Pressure rise} = \frac{62.4 \times 1000 \times 1}{32.2 \times 144} = 13.5 \text{ lb/in.}^2$$

Peak line pressure = 250 + 13.5 = 263.5 lb/in.^{2} gauge

Note velocity of sound in water (acoustic velocity) at stated conditions = 4701 ft/s. Thus we see that, had the valve been closed gradually, the pump connection would not have ruptured.

3.16 Determine vapor pressure of the water by use of steam tables. The vapor pressure of the water at 70°F is found to be $P_v = 0.3631$ lb/in.^{2} abs or 52.3 lb/ft. The water's specific volume from the steam tables is found to be 0.01606 ft^3/lb. Converting this to density, 1/0.01606 = 62.2 lb/ft^3. The vapor pressure in feet of 70°F water is $f_v = 52.3/62.2 = 0.84$ ft of water.

Now determine the friction head loss and the velocity head. From the reservoir to the highest point of the siphon in Exhibit 3.16, the friction head in the pipe must be overcome. Use the Hazen-Williams formula or a pipe friction table to determine the friction head. Use the Hydraulic Institute Pipe Friction Manual, $h_f = 4.59$ ft per 100 ft of pipe or (500/100)(4.59) = 22.95 ft. From the same table, velocity head = 2.63 ft.

Finally, for a siphon handling water, the maximum allowable height, h ft, at sea level with atmospheric pressure at 14.7 lb/in^2 abs is found to be equal to (14.7 × 144 in.^{2}/ft^2 per density of water at operating temperature, lb/ft^3) – (vapor pressure of water at operating temperature, ft + 1.5 × velocity head, ft + friction head, ft). Thus,

h = (14.7 × 144/62.2) – (0.84 + 1.5 × 2.63 + 22.95) = 11.32 ft

In actual practice, the value of h is taken as 0.75 to 0.80 of the computed value. Thus,

11.32 × 0.75 = 8.5 ft.

3.17 Assume a steam velocity of 12,000 ft/min, which is a reasonable one. From steam tables, the specific volume v is found to be 0.7604 ft³/lb. Determine the cross-sectional area as

$$a = \frac{2.4(72,000)(0.7604)}{12,000} = 10.98 \text{ in.}^2$$

$$\text{ID} = 2\left(\frac{a}{\pi}\right)^{1/2} = 3.74 \text{ in. Use a 4-in.-diameter pipe}$$

Determine the pipe schedule number. The ANSA Code for Pressure Piping defines schedule number as SN = 1000 P_1/S. Assume for this pipe and operating conditions that a seamless ferritic alloy steel (1 percent Cr, 0.55 percent Mo) pipe is used.

$$SN = \frac{(1000)(1014.7)}{13,100} = 77.5$$

$$= 1000\frac{P_1}{S}$$

Use the next higher SN, which is 80.

Pipe-wall thickness may be found from piping handbooks. The thickness for a 4-in., SN 80 pipe is 0.337 in.

Note: Use the method given here for any type of pipe—steam, oil, water, gas, or air—in any service—power, refinery, process, commercial, etc. Refer to the proper section of B31.1 Code for Pressure Piping when computing the schedule number, because the allowable stress S varies for different types of service.

3.18 Assume large Reynolds number

Orifice coefficient = $C_0 = 0.61$

Pressure drop = height of liquid

Orifice diameter/tank diameter = D_o/D_T = much less than 1

$$A_o = 0.7854(1/2)^2 = 0.1963 \text{ ft}^2$$
$$\text{Flow} = (0.61)(0.1963)\sqrt{2gh} = 4.30$$
$$A_l = \pi(20)^2/4 = 314 \text{ ft}^2$$

And $314\, dh/dt = -0.958\sqrt{h}\int_o^{20} dh/\sqrt{h}$

$$= -(0.985/314)\int_o^t dt$$

$2\sqrt{h} = -0.00305 + C$. Note when $t = 0, h = 20$. Therefore, $C = 8.94$.

$$t = \frac{-2\sqrt{h} + 8.94}{0.00305} = \frac{8.94}{3.05} = 2920 \text{ s} \quad \text{or} \quad 48.67 \text{ min}$$

Time to empty tank from full level to zero level = about 48.67 min.

3.19 See Exhibit 3.19. And $(P_2/\rho) - Z_1(g/g_c) - h_f = 0$ via Bernoulli. Friction loss through the line must remain constant so long as P_2 and Z_1 do. The Fanning friction factor $f = (h_f\, Dg_c)/(2Lv^2) = $ constant.

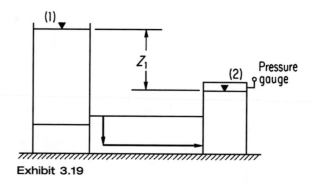

Exhibit 3.19

$$h_f = 2\frac{0.046/R_e^{0.2}}{Dg_c}Lv^2 = \frac{2(0.046)(L)(v^2)}{(D^{0.2}\rho^{0.2}v^{0.2})\rho^{0.2}}D_c$$

$$h_f = \frac{P_2}{\rho} - Z_1\frac{g}{g_c}\,h_f = \frac{0.092Lv^{1.8}\mu^{0.2}}{D^{1.2}\rho^{0.2}g_c}$$

Flow rate $= W = \rho A v$

$$W^{1.8} = \rho^{1.8}A^{1.8}v^{1.8} = \frac{\rho^{1.8}A^{1.8}h_fD^{1.2}\rho^{0.2}g_c}{(0.092)L\mu^{0.5}}$$

$$W^{1.8} = \frac{\rho^{1.8}\pi^{1.8}D^{3.6}}{4^{1.8}}\frac{h_f\rho^{0.2}g_cD^{1.2}}{0.092L\mu^{0.2}}$$

Everything is constant except W and D; hence

$$W^{1.8} = BD^{4.8}$$

where B is a constant.

For the original case, $W_1^{1.8} = BD_1^{4.8}$. Double flow rate $(2W_1)^{1.8} = BD_2^{4.8}$. Taking ratios gives

$$\left(\frac{2W_1}{W_1}\right)^{1.8} = \frac{BD_2^{4.8}}{BD_1^{4.8}} \quad \text{or} \quad \frac{D_2}{D_1} = 2\times\frac{1.8}{4.8} = 2^{0.374} = 1.296 \quad \text{or} \quad 1.3$$

Therefore, doubling pipe diameter is wrong. Doubling cross-sectional area is increasing diameter by $2 = 1.414$, so this is also wrong. Use $D_2 = 1.3D_1$ to double the flow.

3.20 Pressure drop through orifice $= 1.2 \times 1.133 = 1.36$ ft of water

$$Q = A_o C_d (2g\Delta h)^{1/2}, \text{ where } A_o = 0.7854(3/12)^2 = 0.0491 \text{ ft}^2, \text{ and } C_d = 0.61$$

C_d evaluated from $d/D = 3/6 = 0.5$ ratio. Then

$$Q = 0.0491 \times 0.61(64.4 \times 1.36)^{1/2} = 0.2803 \text{ ft}^3\text{/s (actual flow rate)}.$$

Also, $Q = A_p V_p$,   where $A_p = 0.7854 (6/12)^2 = 5148$ ft/h.

$\text{hp} = Q\rho H_t /550$,   where $H_t = H_s + H_f + H_v$,

and, $Q = \text{ft}^3\text{/s}, \rho = 62.27$ lb/ft^3

$H_s = 50$ ft given, but H_f will be evaluated from use of Reynolds

number to find friction factor, so that $H_f = fLV^2/2gD$,

where $L = 500$ ft.

Reynolds number $= VD\rho/\mu = (5184 \times 0.5 \times 62.27)/(2.37) = 6.76 \times 10^4$. At this value for Reynolds number the friction factor for smooth pipes is $f = 0.021$. Then

$$H_f = (0.021 \times 500 \times (1.43)^2)/(64.4 \times 0.5) = 0.67 \text{ ft of water}$$

$$H_v = V^2/2g \text{ (because all the water must be accelerated to the}$$
$$\text{velocity in the orifice).}$$

And $v_0 = Q/A_o = 0.2803/0.0491 = 5.71$ ft/s

$$H_v = (5.71)^2/64.4 = 0.506 \text{ ft of water}$$

Therefore, $H_t = 50 + 0.67 + 0.506 = 51.2$ ft and finally

$$\text{hp} = 0.2803 \times 62.27 \times 51.18 (550 \times 0.78)$$
$$= 2.082 \text{ hp, say, } 2.0 \text{ hp}$$

Heat Transfer

OUTLINE

PROBLEMS

4.1   A 4530-sq-ft heating surface counterflow economizer is used in conjunction with a 150,000-lb per hr boiler. The inlet and outlet water temperatures are 210 and 310°F. The inlet and outlet gas temperatures are 640 and 375°F. Find the overall heat transfer coefficient in Btu per hr per sq ft per °F.

4.2   A surface feed-water heater is to be designed to heat 500,000 lb of water per hr from 200 to 390°F; the heater is to be straight condensing with no desuperheat or subcooling zones. Saturated steam at 400°F is to be used as the heating medium; drains leave as saturated liquid at 400°F. The tubes are $\frac{7}{8}$ in. outside diameter with $\frac{1}{16}$-in. walls. The heater is to have two water passes and it has been estimated that the overall coefficient of heat transfer is 700 Btu/(hr)(°F)(sq ft) of outside tube surface. Specify the number and length of the tubes.

4.3   A company is heating a gas by passing it through a pipe with steam condensing on the outside. It is proposed to triple the capacity of the heater by changing its length while maintaining the same terminal conditions regarding temperature. What percentage change in length is necessary?

4.4   A properly designed steam-heated tubular preheater is heating 45,000 lb per hr of air from 70 to 170°F when using steam at 5 psig. It is proposed to double the rate of airflow through the heater and still heat the air from 70 to 170°F; this to be accomplished by increasing the steam pressure. Calculate the new steam pressure required to meet the changed condition, expressed as psig.

4.5   A counterflow bank of boiler tubes has a total area of 900 sq ft and its U is 13 Btu/(hr)(sq ft)(°F). The boiler tubes generate steam at a pressure of 1000 psia. The tube bank is heated by flue gas, which enters at a temperature of 2000°F and at a rate of 450,000 lb per hr. Assume an average specific heat of 0.25 Btu/(lb)(°F) for the gas and calculate (a) the temperature of the gas that leaves the bank of boiler tubes. Also calculate (b) the rate at which steam is being generated in the tube bank.

4.6 A fuel-oil heater is to be purchased to heat 10,000 gal per hr of an oil initially at 60°F. Exhaust steam at atmospheric pressure from a reciprocating engine is to be used as the heating medium. The general design of the heater has already been decided and it is believed that the overall coefficient of heat transfer will be 12.5 Btu/(hr)(sq ft)(°F) arithmetic mean temperature difference. Total installed cost of the heater will be $3 per sq ft. Annual fixed charges are to be figured at 40 percent; maintenance and repairs are to be considered negligible. Heat added to the oil is valued at 40 cents per million Btu. No charge is to be placed against the heating steam because it would be wasted to atmosphere if not used in the heater. The heater is to operate 3000 hr per yr. The oil has a constant specific heat of 0.5. Its average specific gravity during passage through the heater is to be taken as 0.9.

(a) To what economical temperature should the oil be heated? No differential in pumping charge is to be assumed.

(b) What heating surface is to be associated with that economical temperature?

4.7 A water-to-air heat exchanger operates under the following conditions:

Air mass flow rate............................ 701,400 lb per hr
Air specific heat, constant.................... 0.240
Air entering temperature..................... 60°F
Air leaving temperature..................... 171°F
Water mass flow rate......................... 286,000 lb per hr
Water specific heat........................... 1.0
Water entering temperature................. 203°F

The exchanger is to be used to heat the same air quantity from minus 10°F to 121°F with water entering at 203°F. Based on the assumption that the U in Btu/(hr)(sq ft)(°F) log mean temperature difference is unchanged, determine the required mass flow rate of water for these new conditions.

4.8 A hot-water heater consists of a 20-ft length of copper pipe of $\frac{1}{2}$-in. average diameter and $\frac{1}{16}$-in. thickness. The outer surface of the pipe is maintained at 212°F. What is the capacity of the coil in gpm if water is fed into the coil at 40°F and is expected to emerge heated to 150°F? Assume the conductivity of the copper to be 2100 Btu/(sq ft)(°F)(hr) per in. of thickness.

4.9 A steam condenser receives steam at 1.5 in. Hg absolute and a quality of 0.96. Water "on" at 65°F, "off" at 82°F. Condensate leaves at 86°F, condensing surface 40,000 sq ft, condensing rate 30,000 lb per hr. Determine (a) overall coefficient of heat transfer and (b) circulating water required in gpm.

4.10 Repairs are to be made on a valve in a steam line. When the valve is isolated, the insulation is removed from the valve and the valve surface temperature is measured and found to be 700°F with an ambient temperature of 60°F.

Thirty minutes later the valve surface temperature is again measured and found to be 500°F with an ambient temperature still 60°F. Estimate the time required for the valve surface temperature to reach 150°F, assuming that the ambient temperature remains constant.

4.11 A concrete storage tank in the shape of a cube and lined with steel plate with no air space between steel and concrete is to hold 7500 gal of a sodium hydroxide solution. Walls are 12 in. thick. (a) How much heat in Btuh must be added to the solution to prevent it from dropping below 25°F if the outside temperature is 0°F? (b) How many feet of 1-in. steel pipe would you install as a heating coil, using 5 psig saturated steam to maintain the proper temperature?

4.12 A certain clean, metal, double-pipe heat exchanger heats turbulently flowing air to a particular temperature by means of condensing steam. How much greater an air rate should we be able to process with two such heat exchangers in series, if steam pressure and terminal temperatures remain the same?

4.13 Stack gases from a chemical processing unit operation are composed primarily of air, with small amounts of noxious vapors that must be condensed out to prevent a community air pollution problem. The gases are flowing at a rate of 80,000 cu ft per hr, leaving the process at a temperature of 600°F, and are to be cooled down to 100°F. Cooling water is available at 50°F to cool the tube side of a shell-and-tube heat exchanger of cross-flow design. Gases will flow on the shell side of the exchanger. Estimate the surface required to do the job.

4.14 If 10,000 lb per hr of distilled water is required from untreated water, steam is available at 300°F, and the condenser will vent to the atmosphere, how much surface is required? Assume a pressure drop through the condenser and lines of about 5 psi. The saturation temperature in the evaporator shell will be 19.7 psia or 226°F. Take the overall coefficient = 605.

4.15 A fixed-displacement pump supplies fluid to two hydraulic motors driving a rolling mill. The speed of the motor is accurately controlled by two pressure-compensated flow-control valves used in a meter in the circuit. The average controlled flow is 20 gpm, the flow from the pump is 25 gpm, and the system pressure is 2500 psi. Heat is generated by the pumping of the superfluous 5 gpm through the relief valve, but the extra capacity this represents must be maintained to handle the peak loads. The following data apply:

1. The oil temperature rise by natural convection from the pump reservoir is 137°F.

2. It is desired to hold the reservoir temperature to 90°F; a 40°F rise in fluid temperature (to 130°F) is permissible.

3. Cooling water from the spray towers has a summer temperature of 85°F.

4. A water flow of 10 gpm is available and seems reasonable.

5. The overall coefficient of heat transfer oil-to-water is 100 Btu/(hr)(sq ft)(°F).

Find the surface required for the oil cooler to hold the reservoir temperature at 90°F.

4.16 A kiln is to be designed for our plant to withstand a temperature of 2000°F while limiting the heat loss to 250 Btu/(h)(ft²) with an outside temperature of 100°F. We have available the following types of brick:

Fireclay	4.5 in. thick		$k = 0.90$
Insulating	3.0 in. thick	1800°F max allowable temp	$k = 0.12$
Building	4.0 in. thick	300°F max allowable temp	$k = 0.40$

   a. What will be the minimum wall thickness?
   b. Determine the actual heat loss through the wall.

4.17 A furnace is constructed with 9 in. of firebrick, $4\frac{1}{2}$ in. of insulating brick, and 9 in. of building brick. The inside surface temperature is 1400°F and the outside air temperature is 85°F. If the furnace wall loses heat at the rate of 200 Btu/(h)(ft²), determine the air heat-transfer coefficient.

SOLUTIONS

4.1 Refer to Exhibit 4.1. Make use of the basic heat transfer equation: $Q = UA\,\Delta t_m$, $U = Q/(A\,\Delta t_m)$. The amount of heat transferred Q is obtained from the familiar expression $Q = wc_p\,\Delta t$.

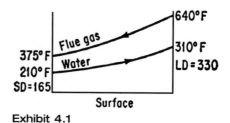

Exhibit 4.1

$$Q = 150,000 \times 1 \times (310 - 210) = 15,000,000 \text{ Btu per hr}$$

From the formula for log mean temperature difference determine Δt_m.

$$\Delta t_m = \frac{330 - 165}{\ln\left(\frac{330}{165}\right)} = 238$$

We now know all factors in the basic equation so that

$$U = \frac{Q}{A\,\Delta t_m} = \frac{15 \times 10^6}{4530 \times 238} = 13.9 \text{ Btu/(hr)(sq ft)(°F)}.$$

4.2 Refer to Exhibit 4.2 for the heat-transfer process. The log mean temperature difference may be determined by direct calculation, and is found to be 64°F. The total amount of heat transferred may be found from the water side, assuming complete insulation of the shell.

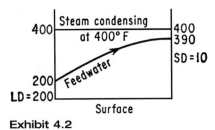

Exhibit 4.2

$$Q = 500,000 \times 1 \times (390 - 200) = 95 \times 10^6 \text{ Btu per hr}$$

The surface on the heater may also be found to be equal to

$$A = \frac{Q}{U \, \Delta t_m} = \frac{95 \times 10^6}{700 \times 64} = 2120 \text{ sq ft.}$$

The number of tubes per pass is simply determined by first finding the flow rate in cubic feet per second, then the total cross-sectional area of the tubes and finally dividing by the internal diameter of each tube.

$$\text{Tubes per pass} = \frac{\text{cu ft per sec}}{6 \text{ fps}} \times \frac{1}{0.00307 \text{ sq ft}}$$

$$\text{Tubes per pass} = \frac{500,000/(62.4 \times 3600)}{6} \times \frac{1}{0.00307} = 121 \text{ tubes}$$

The velocity of 6 fps for water flow speed is an acceptable value and is often used in heat-exchanger design. Now for a two-pass heater the number of tubes total is $2 \times 121 = 242$ tubes. The effective length of tubes is

$$\frac{2120}{A_0 \times 242} = \frac{2120}{0.2291 \times 242} = 38.2 \text{ ft}$$

The outside area per foot length of tube A_0 is 0.2291 sq ft.

4.3   Present capacity is $Q_1 = UA_1 \Delta t_m$. New capacity will be

$$Q_2 = U3A_1 \Delta t_m = 3Q_1.$$

Therefore, increase in length is 200 percent. Note that area is equal to diameter × length.

4.4   Refer to Exhibit 4.4. Under present conditions the heat transferred is

$$Q_1 = 45,000 \times 0.25 \times (170 - 70) = 1,125,000 \text{ Btu per hr.}$$

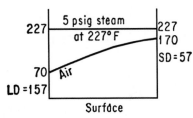

Exhibit 4.4

Under the new conditions the heat to be transferred is

$$Q_2 = 2 \times Q_1 = 2 \times 1{,}125{,}000 = 2{,}250{,}000 \text{ Btu per hr.}$$

The only change is an increase in Δt_m, with U and A remaining the same. From Exhibit 4.4, the log mean temperature difference is 98, say 100°F. Under the new conditions the Δt_m would be 200°F. The net effect will be to raise the horizontal steam line in Exhibit 4.4. As this happens we can safely say that the arithmetic mean is permissible. Then

$$\frac{LD + SD}{2} = 200 \quad LD + SD = 400$$

$$LD = x - 70 \quad SD = x - 170 \quad (x - 70) + (x - 170) = 400$$

$$x - 70 + x - 170 = 400 \quad x = \frac{400 + 70 + 170}{2} = \frac{640}{2} = 320°F$$

The saturated steam pressure corresponding to 320°F may be obtained from the steam tables as 89.65 psia, or

$$89.65 - 14.7 = 75 \text{ psig}$$

4.5 The basic heat-transfer equation as indicated below applies:

$$Q = w_g c_p \Delta t_g = UA \, \Delta t_m.$$

Now refer to Exhibit 4.5. Steam being generated in the tube bank and steam line is below the gas line. From the steam tables the steam temperature corresponding to a saturation pressure of 1000 psia is 544.56°F. At this condition the enthalpy of the liquid H_f is 542.4, of vaporization H_{fg} is 649.5, and steam enthalpy H_g is equal to 1191.9.

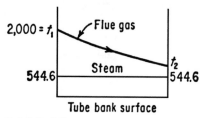

Exhibit 4.5

We know all terms to be inserted into the basic heat transfer equations except the leaving gas temperature. Let us call this t_2 (see Exhibit 4.5). Then

$$Q = 450{,}000 \times 0.25 \times (2000 - t_2) = 13 \times 900 \times \Delta t_m$$

$$\Delta t_m = \frac{2000 - t_2}{\ln[(2000 - 544.6)/(t_2 - 544.6)]}.$$

By rearrangement,

$$\ln \frac{2000 - 544.6}{t_2 - 544.6} = \frac{13 \times 900}{450{,}000 \times 0.25} = 0.104.$$

The antilog of 0.104 = 1.11.

a. Leaving gas temperature:

$$\frac{2000 - 544.6}{t_2 - 544.6} = 1.11$$

from which $t_2 = 1850°F$.

b. Steam-generation rate: By simple heat balance the heat absorbed by the water to generate steam is equal to the heat lost by the flue gas.

$$Q = w_g c_p(2000 - t_2) = 450,000 \times 0.25 \times (2000 - 1850) = 16.9 \times 10^6$$

If we assume water enters tube bank saturated at 544.6°F, then

Steaming rate $\times H_{fg} = 16,900,000$ Btu per hr.

Finally,

Steaming rate $= 16,900,000/649.5 = 26,000$ lb steam per hr.

4.6 Refer to Exhibit 4.6. The weight of oil to be heated is simply

$$w_0 = 10,000 \times 8.33 \times 0.90 = 75,000 \text{ lb per hr.}$$

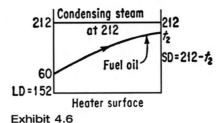

Exhibit 4.6

The heat transferred to the oil to raise it from 60°F to t_2 is also simply

$$Q = w_0 c_p(t_2 - t_1) = 75,000 \times 0.5(t_2 - 60) \text{ Btu per hr.}$$

Annual fixed charges (FC) based on the heater surface A in square feet are

$$FC = \$3 \times A \times 0.40 = \$1.20\,A.$$

Ordinarily the steam would be wasted to the atmosphere, but it is used to heat the oil and is thus saved. This saving expressed in terms of leaving oil temperature is so many dollars worth.

$$\begin{aligned} S &= Q \times 3000 \times 0.40 \text{ dollars per million Btu} \\ &= 75,000 \times 0.5(t_2 - 60) \times 3000 \times 0.40 = (45t_2 - 2700) \end{aligned}$$

The fixed charges are to balance out the saving. We now have two equations with two unknowns. Let us solve for the heater surface A. We have been given the overall coefficient U as 12.5. For the conditions of the problem setup we can safely assume the arithmetic mean temperature difference. Thus,

$$U = \frac{Q}{A \times \Delta t_a} \quad A = \frac{Q}{U \times At_a} = \frac{37,500(t_2 - 60)}{12.5 \times \Delta t_a}$$

$$\Delta t_a = \frac{LD + SD}{2} = \frac{(152 - 60) + (212 - t_2)}{2} = \frac{364 - t_2}{2}$$

$$FC = 1.2A = 1.2 \times \frac{37,500(t_2 - 60)}{12.5 \Delta t_a} = \frac{3600(t_2 - 60)}{\Delta t_a}$$

$$= \frac{3600(t_2 - 60)}{\Delta t_a} = \frac{7200(t_2 - 60)}{364 - t_2}$$

Equating S equal to FC, we have

$$45t_2 - 2700 = \frac{7200(t_2 - 60)}{364 - t_2}$$

$$45t_2 - 2700 = \frac{7200t_2 - 432,000}{364 - t_2}.$$

This may be simplified to the binomial equation

$$t_2^2 - 264t_2 - 12,280 = 0$$

a.   from which $t_2 = 206°F$.

b.   heating surface $A = \dfrac{37,500(206 - 60)}{12.5[(364 - 206)/2]} = 5520$ sq ft.

4.7   First determine the overall coefficient U. This will remain unchanged for both loading conditions. Also the product of $U \times A$ will remain constant. Let us denote by subscript 1 the first set of conditions and by subscript 2 the second set of conditions. Also subscript w is for water.

$$Q_1 = 701,400 \times 0.24 \times (171 - 60) = 286,000 \times 1 \times \Delta t_w$$

$$\Delta t_w = \frac{701,400 \times 0.24 \times 111}{286,000 \times 1} = 65.5°F, \text{ say } 66°F$$

Now set up Exhibit 4.7a for the first condition. Note water leaving temperature is $203 - 66 = 137°F$. By calculation in the usual manner the log mean temperature difference for this set of conditions is $51°F$. If Q_2 is the new air loading ($-10°F$ to $121°F$), then we can say

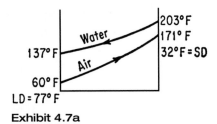

Exhibit 4.7a

$$Q_1 = UA\ \Delta t_{m1} = C\ \Delta t_{m1}$$
$$Q_2 = UA\ \Delta t_{m2} = C\ \Delta t_{m2}$$
$$Q_1 = 701,400 \times 0.24 \times (171 - 60) = 18.6 \times 10^6\ \text{Btu per hr}$$
$$Q_2 = 701,400 \times 0.24 \times (121 + 10) = 22.0 \times 10^6\ \text{Btu per hr}$$

also

$$\frac{Q_1}{Q_2} = \frac{C\ \Delta t_{m_1}}{C\ \Delta t_{m_2}} = \frac{18.6}{22.0} = 0.85 = \frac{51}{\Delta t_{m_2}}$$

$$\Delta t_{m_2} = 51/0.85 = 60°F$$

Now set up Exhibit 4.7b for the second condition. Again by calculation log mean temperature difference of 60°F, and large difference of 203 − 121 = 82°F, the small difference is found to be 42°F. This makes the leaving water temperature 42 − 10 = 32°F. This is the freezing temperature of water, indeed, but no freeze-up will occur so long as the water is flowing. For best practical operation of the exchanger, a control temperature of 35°F should be used. Finally,

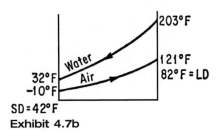

Exhibit 4.7b

$$Q_2 = 22 \times 10^6 = \text{lb water per hr} \times c_p \times \Delta t_{w_2}$$

$$\text{lb water per hr} = \frac{22 \times 10^6}{1 \times 171} = 128,500$$

or the equivalent of

$$128,500/(60 \times 8.33) = 257\ \text{gpm.}$$

4.8   Assuming clean pipe and no fouling,

$$\frac{1}{U} = R, \text{ steam film} + R, \text{ pipe wall} + R, \text{ water film}$$

$$\frac{1}{U} = \frac{1}{1000} + \frac{0.0625}{2100} + \frac{1}{550} = 0.0028298$$

$$U = 1/0.0028298 = 354 \text{ Btu/(hr)(sq ft)(°F)}$$

$$A = 20 \times \pi \times 0.5/12 = 2.618 \text{ sq ft active surface}$$

$$212 \rightarrow 212$$

$$\underline{150} \leftarrow \underline{40}$$

$$\text{SD} = 62 \qquad 172 = \text{LD}$$

log MTD = 180°F. This is uncorrected for the type of flow. Further refinement is not required. Note also that U was determined for thin-wall tubing with no correction for wall thickness. The resistance due to wall is negligible. Now,

$$Q = UA \, \Delta t_m = 354 \times 2.618 \times 108 = 100{,}091 \text{ Btu per hr.}$$

Assuming that all heat goes into heating up the water (complete insulation),

$$Q = w c_p \Delta t = 100{,}091 = w \times 1 \times (150 - 40)$$

$$w = 100{,}091/110 = 910 \text{ lb per hr,} \quad \text{or} \quad \frac{910}{500} = 1.82 \text{ gpm.}$$

4.9   Use Mollier diagram and steam tables. Assume no radiation losses. Also refer to Exhibit 4.9. Let the following nomenclature apply:

Q = heat added to condenser water, Btu per lb
Q = heat given up by steam, Btu per lb
c_p = specific heat of water, Btu/(hr)(°F)
t_{w_1} = entering water temperature, 65°F
t_{w_2} = leaving water temperature, 82°F
t_{s_1} = entering steam temperature, 92°F
t_{s_2} = leaving condensate temperature, 86°F
t_m = log mean temperature difference, °F

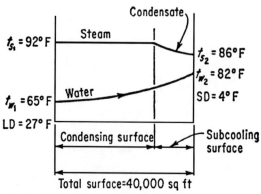

Exhibit 4.9

H_2 = enthalpy of entering steam, 1060 Btu per lb
H_3 = enthalpy of condensate, 54 Btu per lb
A = condensing surface, 40,000 sq ft
U = overall heat-transfer coefficient, Btu/(hr)(sq ft)(°F)

a. Overall coefficient of heat transfer:

$$Q = 30,000(H_2 - H_3) = 30,000(1060 - 54) = 30.1 \times 10^6 \text{ Btu per hr}$$

Refer to Exhibit 4.9 and determine log MTD to be 12°F. Then,

$$U = \frac{Q}{A\,\Delta t_m} = \frac{30.1 \times 10^6}{40,000 \times 12} = 63 \text{ Btu/(hr)(sq ft)(°F)}.$$

b. Cooling water:

$$w = \frac{30.1 \times 10^6}{82 - 65} = 1.772 \times 10^6 \text{ lb per hr}$$

$$\frac{1.772 \times 10^6}{500 \times \text{sp gr}} = \frac{1.772 \times 10^6}{500 \times 1} = 3540 \text{ gpm}.$$

In certain problems presented by several state boards, conductivity of metal wall has been presented as $27 - t/180$ Btu/(hr)(sq ft)(°F)(ft) thickness for steel. Here t is the arithmetic average of the two temperatures across the wall having a flat surface (not rounded).

4.10 For more accurate results three initial surface temperatures should be taken and then plotted so that temperature is taken along the abscissa in logarithm form and time is plotted along the ordinate in cartesian form as in Exhibit 4.10.

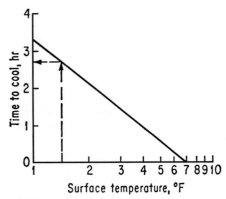

Exhibit 4.10

From the plot in Exhibit 4.10 the time is approximately 2 hr, 75 min, say 3 hr. The greatest change takes place after the first reading, then is linear thereafter.

4.11 a. Using the outside area of the cube and the temperature difference of 25°F and the overall coefficient, which we shall calculate, we can find the heat loss under steady state conditions. By familiar method we find the cube to be 10 ft on a side. If we assume the tank is sitting on the ground, the exposed surface totals 500 sq ft. Determine the coefficient U by getting the summation of all resistances: air film, concrete, steel plate, and sodium hydroxide liquid film, and dividing into unity. $R_t = R_1 + R_c + R_s + R_2$, where

$$R_1 = \frac{1}{6} = 0.167 \quad R_c = \frac{L_c}{k_c} = \frac{12}{8.3} = 1.45$$

$$R_s = \frac{L_s}{k_s} = \frac{1}{35 \times 12} = 0.00238 \quad \text{Note 1-in. steel plate.}$$

$$R_2 = \frac{1}{500} = 0.0020 \quad R_t = 0.167 + 1.45 + 0.00238 + 0.0020$$

From which R_t is 1.621. Now $U = 1/R_t = 1/1.621 = 0.618$ Btuh/(sq ft)(°F).
$$Q = 0.618 \times 500 \times 25 = 7720 \text{ Btuh.}$$

b. Heat transfer coefficient for pipe in sodium hydroxide solution

$$U_p = \frac{1}{\frac{1}{600} + 0 + 0.133/(35 \times 12) + 0(0.344/0.275) + 1/(1000 \times 0.275/0.344)}$$

$$U_p = \frac{1}{0.00167 + 0 + 0.0003165 + 0 + 0.00125} = 308.5 \text{ Btuh/(sq ft)(°F)}$$

Btuh/ft of pipe $= U_p \times A_0 \times (227 - 25) = 308.5 \times 0.344 \times 202 = 21{,}400$

Length of pipe required $= 7720/21{,}400 = 0.36$ ft. For all practical purposes this is an unwieldly length. The best way is to run pipe around the perimeter 1 ft away from the sides of the tank 6 in. off the bottom. This will require about 36 ft of pipe.

4.12 In the exchanger described, virtually all the resistance to heat transfer would be in the air film. We can say, therefore, $U = h$. Substituting this in the usual $Q = UA\,\Delta t_m$ we have

$$Wc_p\,\Delta t = h(\pi DL)\Delta t_m$$

Because c_p, Δt, D and Δt_m are the same for one exchanger as for the two together, we can simplify the preceding equation to $W\alpha hL$. Now, the Dittus-Boelter equation states $Nu = 0.023\,Re^{0.8}\,Pr^{0.4}$. From the Dittus-Boelter equation, we can say $h\alpha W^{0.8}$ because diameter and physical properties are also constant in our problem. Then by substitution from equations substituting $h\alpha W^{0.8}$ in $W\alpha hL$, we get $W\alpha W^{0.8}L$ or $W^{0.2}\alpha L$. Also $W\alpha L^5$ or $W_2/W_1 = (L_2/L_1)^5$. Thus, the flow rate ratio equals the length ratio raised to the fifth power. For the numbers in our problem

$$\frac{W_2}{W_1} = \left(\frac{2}{1}\right)^5 = 32.$$

The two exchangers in series can process 32 times as much air as the single exchanger. Of course, before actually trying this doubling up of heat exchangers, we should calculate the increase in pressure drop. Taking as an approximate friction factor $f \alpha W^{-1/5}$, we have from the Fanning equation

$$\Delta P = \alpha W^{-1/5} W^2 L: \quad \text{or} \quad \Delta P \alpha L^{-1} L^{10} L \quad \text{or} \quad \Delta P \alpha L^{10}.$$

For a doubled length

$$\frac{\Delta P_2}{\Delta P_1} = \left(\frac{2}{1}\right)^{10} = 1024.$$

This is more than a thousandfold increase in the exchanger pressure drop. However, if we only wish to double the air rate through the exchanger, rather than multiply it by 32, then

$$\frac{L_2}{L_1} = \left(\frac{W_2}{W_1}\right)^{0.2} = 2^{0.2} = 1.15.$$

Finally

$$\frac{\Delta P_2}{\Delta P_1} = \left(\frac{L_2}{L_1}\right)^{10} = \left(\frac{W_2}{W_1}\right)^2 = \left(\frac{2}{1}\right)^2 = 4.$$

Thus, to double the air rate we need 15 percent greater length for our heat exchanger and must suffer four times the pressure drop through it.

4.13 Because the surface area is to be estimated, certain assumptions will be made to simplify the solution.

Assumptions:

Cooling water temperature rise = 25°F
Specific heat of stack gases (mostly air) = 0.24
Stack gases are at atmospheric pressure, 14.7 psia
Average stack gas temperature is arithmetic
Molecular weight of stack gases = 29
No fouling factor on shell or tube sides of heat exchanger
Tubes considered thin-walled

Average gas temperature = (600 + 100)/2 = 350°F or 810°R. Gas density is found by

$$\left(\frac{29}{279}\right) \times (460 + 60)/(460 + 350) = 0.049 \text{ lb per cu ft.}$$

Cooling load to be charged to exchanger:

$$80,000 \times 0.049 \times 0.24 \times (600 - 100) = 470,000 \text{ Btuh} = Q.$$

Refer to Exhibit 4.13. Then since $Q = UA\,\Delta t_m$, we must obtain Δt_m and U, then solve for A, surface area in square feet.

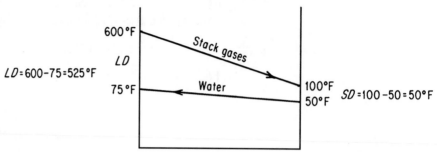

Exhibit 4.13

Log mean temperature difference $\Delta t_m = \dfrac{LD - SD}{2.3 \log (LD/SD)}$

$$\Delta t_m = \frac{525 - 50}{2.3\left(\frac{525}{50}\right)} = 203°\text{F}$$

$$U = \frac{1}{(1/h_o) + (1/h_i)} = \frac{1}{(1/250) + (1/500)} = 167 \text{ Btuh/(°F)(hr)}$$

$$A = 470,400/(167 \times 203) = 14 \text{ sq ft.}$$

4.14 Heat balance is

$$Q_{\text{evap}} = 10,000 \times 961 = 9,610,000 \text{ Btuh}$$
$$Q_{300F} = 10,550 \times 910 = 9,610,000 \text{ Btuh}$$
$$\text{Temperature head } \Delta t = 300 - 226 = 74°\text{F}$$

$$\text{Surface area} = \frac{Q}{U\,\Delta t} = \frac{9,610,000}{605 \times 74} = 215 \text{ sq ft.}$$

4.15 Calculate heat load Q from the following equation:

$$Q = 1.484 \times R \times P.$$

Thus,

$$Q = 1.484 \times 5 \text{ gpm} \times 2500 \text{ psi} = 18,550 \text{ Btu per hr.}$$

Temperature rise of cooling water ΔT_w is given by

$$\Delta T_w = 18,550/(60 \times 10 \times 8.33) = 3.725°\text{F.}$$

Thus, cooling water can be assumed to enter at 85°F and leave at 89°F. The log mean temperature difference LMTD is now found as follows:

$$\text{LMTD} = \frac{\Delta T_{\text{max}} - \Delta T_{\text{min}}}{2.3026 \log_{10}(\Delta T_{\text{max}}/\Delta T_{\text{min}})} = \frac{(130 - 85) - (90 - 89)}{2.3026 \log_{10}(130 - 85)/(90 - 89)}$$
$$\text{LMTD} = 11.6°\text{F.}$$

Finally, the surface required is found.

$$A = Q/(U \times \text{LMTD}) = 18{,}550/(100 \times 11.6) = 16 \text{ sq ft}$$

If an exchanger having this heat transfer surface is not available, the next larger size should be selected. It is advisable to check the water pressure drop to make certain it is within acceptable limits.

4.16 Refer to Exhibit 4.16. x_1, x_2, and x_3 must be multiples of 4.5, 3, and 4 in., respectively.

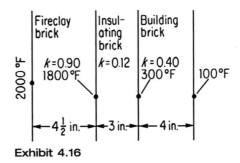

Exhibit 4.16

a.   Basis: 1 ft^2 of surface area. Let us use the series formula

$$Q = \Delta t/\Sigma R, \quad R = x/kA, \quad A = 1 \text{ ft}^2$$
$$Q = 250 \text{ Btu/(h)(ft}^2)(1 \text{ ft}^2) = 250 \text{ Btu/h.}$$

For fireclay, $t_a \geq (2000 - 1800) = 200°F$.

$$R = (x_1/k_1A) = \Delta t_1/Q \quad \text{Also } x_1 = k_1A(\Delta t_1/Q)$$
$$x_1 = 0.9 \ (1 \text{ ft}^2)(200/250) = 0.72 \text{ ft or } 8.64 \text{ in.}$$

Thus we need two layers of firebrick. $x_1 = (2)(4.5) = 9$ in. or 0.75 ft
Then $\Delta t_1 = (x_1)/(k_1A)(Q) = (0.75)(250)/(0.9) = 208°F$
Now $\Delta t_2 \geq 2000°F - 208°F - 300°F = 1492°F$

$$x_2 = 0.12(1 \text{ ft}^2)(1492/250) = 0.716 \text{ ft or } 8.59 \text{ in.}$$

Thus we need three layers of insulating brick. $x_2 = (3)(3 \text{ in.}) = 9$ in. or 0.75 ft

Then $\Delta t_2 = (x_2)/(k_2A)(Q) = (0.75)(250)/(0.12) = 1563°F$
Now $\Delta t_3 \geq (2000°F - 208 - 1563 - 100) = 129°F$
$$x_3 = 0.4(1)(129/250) = 0.206 \text{ ft or } 2.48 \text{ in.}$$

Thus we need one layer of building brick. $x_3 = 1(4 \text{ in.}) = 0.33$ ft

Now the minimum wall thickness is 9 in. + 9 in. + 4 in. = 22 in.

$$Q = \frac{\Delta t}{\Sigma R} = \frac{2000 - 100}{(0.75/0.9) + (0.75/0.12) + (0.333/0.4)}$$

$$= (1900/(7.916) = 240 \text{ Btu/(h)(ft}^2)$$

4.17

$$\frac{Q}{A} = \frac{k_1(t_1 - t_2)}{l_1} = \frac{k_2(t_2 - t_3)}{l_2} = \frac{k_3(t_3 - t_4)}{l_3} = h(t_4 - t_5)$$

Overall temperature drop across wall $= t_1 - t_4$

$$\Delta t = \frac{Q}{A}\left(\frac{l_1}{k_1} + \frac{l_2}{k_2} + \frac{l_3}{k_3}\right) = 200\left[\frac{9}{12(0.8)} + \frac{4.5}{12(0.15)} + \frac{9}{12(0.4)}\right]$$

$$= \frac{200}{12}(11.25 + 30 + 22.5) = 1063°F$$

$$1400 - t_4 = 1063°F \qquad h(337 - 85) = 200$$

$$t_4 = 337°F \quad \text{and} \quad h = 0.794 \text{ Btu/(h)(ft}^2\text{)(°F)}$$

Machine Design

OUTLINE

PROBLEMS

5.1   Calculate the torsional stress in the shaft of a blower from the following data: shaft diameter 2 in., speed 200 rpm, total head of air equivalent to 2 in. WG, volume discharged 30,000 cfm, and blower efficiency 60 percent.

5.2   A truck body lowers 5 in. when a load of 25,000 lb is placed on it. What is the frequency of vibration of the truck if the total spring-borne weight is 40,000 lb?

5.3   A mass weighing 25 lb falls a distance of 5 ft on the top of a helical spring having a spring constant k of 20 lb per in. (a) What will be the velocity of the mass after it has compressed the spring 8 in.? (b) Calculate the maximum compression of the spring.

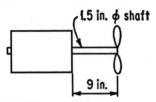

Exhibit 5.4

5.4   Investigate the acceptability of the fan and motor drive shown in Exhibit 5.4 from the standpoint of possible excessive vibration. Motor: a-c 110 volts, 60 cycles, 2400 rpm. Armature: 40 lb. Radius of gyration: 5 in. Three-blended fan: wt 10 lb; radius of gyration 9 in. Shaft is of steel. If design is not acceptable, indicate what changes should be made.

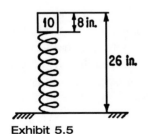

Exhibit 5.5

5.5   A 22-coil squared and ground spring is designed to fire a 10-lb projectile into the air. The spring has a 6-in. diameter coil with $\frac{3}{4}$-in. diameter wire; it has a free length of 26 in. and is compressed 8 in. when loaded or set. The shear elastic limit for the spring material is 85,000 psi. The spring constant Wahl factor is 1.18. The shear modulus of elasticity of the spring material is 12,000,000. (a) Determine the height to which the projectile will be fired. (b) Determine the safety factor for this spring (see Exhibit 5.5).

5.6   A rod shown in Exhibit 5.6 forms a circular path ABC for a slider S. It is connected to A by a spring normally 2.5 ft long with a spring constant of

1 in. per lb force applied. The block is held at C and then released. What work has been done on the block when it reaches B, midway between C and A?

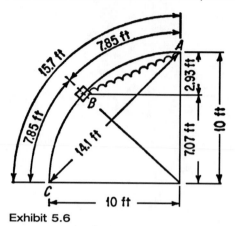

Exhibit 5.6

5.7   It is known that a certain piece of machinery weighing 2 tons creates a disturbing force of 1200 lb at a frequency of 2000 cpm. The machine is to be supported on six springs, each taking an equal share of the load in such a way that the force transmitted to the building is not to exceed 20 lb. Assume that the machine is guided so that it may move only in a vertical direction. Calculate the required scale of each spring.

5.8   An automobile has main helical springs that are compressed 6 in. by the weight of the car body. If the axles of the wheels of the auto are clamped to a test platform and the platform is given a vertical harmonic motion having an amplitude of 1 in. and a frequency of 1 cps, determine the amplitude of the motion of the body of the car. Assume that there are no shock absorbers and thus the vibration takes place without damping. What is the maximum shortening of the spring?

5.9   A light artillery gun is mounted on a wheeled carriage; combined weight of gun and carriage is 3000 lb. The gun fires a 60-lb projectile, which has a velocity of 1500 fps just as it leaves the gun. The angle at which the projectile is fired is 30° above the horizontal. Neglecting the recoil due to powder acceleration, determine the velocity of the gun and carriage just after the projectile leaves the gun.

5.10 The impeller of a slow-speed centrifugal pump has a mass moment of inertia of 5 in.-lb-sec^2. The pump is driven by an electric motor that has a mass moment of inertia of 10 in.-lb-sec^2. A steel shaft whose diameter is 1 in. and whose length is 12 in. connects pump and motor. At a constant speed of 120 rpm, 5 hp are transmitted through the shaft with no measurable vibration. A steel rod enters the impeller eye along with the liquid and jams the impeller, causing the rotor to stop rotating instantly. At the same time a circuit breaker opens the electrical circuit to the motor and the motor stops instantly. What is the resulting maximum shear stress in the shaft?

5.11 It has been estimated that 4000 ft-lb of energy must be supplied at the drive shaft of a belt-driven punch press to blank satisfactorily a hole in a sheet of 115-lb coke tin. Of the total energy required, 800 ft-lb are supplied by the belt and the remainder by the flywheel. The weight of the flywheel must be such that its speed will not vary more than 10 percent from its maximum speed of 200 rpm. Assuming the radius of gyration to be 18 in., calculate the weight of the flywheel.

5.12 One of the routine steps in preliminary design of a piece of turbomachinery is to compare the critical and operating speeds. Such a routine check disclosed that a centrifugal fan would be operating at 98 percent of its first critical speed.
   a. What would be the effect on critical speed caused by decreasing bearing span 10 percent?
   b. What would be the effect on critical speed caused by increasing shaft diameter 10 percent?
   c. What would be the effect on critical speed caused by increasing rotor weight 10 percent?

   Approximate quantitative results are desired.

5.13 A motor weighing 100 lb is mounted on four springs, each having a spring constant k equal to 15 lb per in. What is the natural frequency and period of the motor as it vibrates?

5.14 What is the natural frequency of a shaft of weight W simply supported and of length l? How can this approach be applied to similar systems?

5.15 A variable-speed motor is mounted on an elastic beam at the center of the beam span. The motor weighs 23 lb and drives a shaft that has a 2.0-lb eccentric weight located 1.0 ft from the shaft center. When the motor is not running, the combined motor and eccentric weight causes a 0.50-in. deflection in the supporting beam. (a) Determine the speed of the system at resonance. (b) Determine the amplitude of the forced vibration when the motor is running at 360 rpm.

5.16 A heavy-duty spring is composed of two concentric coils (Exhibit 5.16a). The outside coil, with an outside diameter of 9 in., has six active coils of $1\frac{1}{2}$-in.-diameter round bar stock. The inner spring has nine active coils of 1-in. round bar stock. The outside diameter of the inner coils is $5\frac{1}{2}$ in. The free height of the outer spring is $\frac{3}{4}$ in. greater than that of the inner spring. For a total load of 20,000 lb, find the deflection of each spring and the load carried by inner and outer coils.

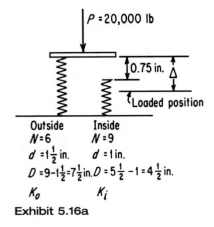

Exhibit 5.16a

5.17 A flexible coupling has an active element in the form of a torsional helical spring. At 900 rpm, 25 hp is transmitted. To what radius of helix must the spring be wound if it is made of $\frac{3}{4}$-in. square wire? If the permissible stress is 30,000 psi, how many active coils must there be if the torsional deflection is not to exceed 5 degrees?

SOLUTIONS

5.1 First determine the shaft horsepower, then the torque, and finally the stress.

Shaft horsepower = $(30,000/13.8 \times \frac{2}{12} \times 62.4/0.075)/(33,000 \times 0.6) = 15.8$

where specific volume of air is 13.8 cu ft per lb and the term 62.4/0.075 converts feet of water to feet of air, the fluid flowing.

Torque = $T = (15.8 \times 33,000)/(2\pi \times 200) = 415$ lb-ft, or 4,980 lb-in.

Then,

$$\text{Stress} = S_s = \frac{T}{Z_p} = \frac{4980}{\pi \times 0.5} = 3180 \text{ psi.}$$

5.2 The frequency is $1/T = W/2\pi$ and the acceleration of the weight is

$$a = -W^2 x$$

but $F = Ma$ and the force acting on the mass is $-kx$.

Then $$-kx = -W^2 Mx$$

and $$W = \sqrt{\frac{k}{M}} = \sqrt{\frac{kg}{W}}.$$

Now we see that frequency f is determined from the following:

$$f = \frac{W}{2\pi} = \frac{1}{2\pi} \times \sqrt{\frac{kg}{W}}$$

where k is the spring constant and is equal to 25,000/5 in. = 5000 lb per in.

$$f = \frac{1}{2\pi} \sqrt{\frac{5,000 \times 12 \times 32.2}{40,000}} = 1.105 \text{ oscillations per sec.}$$

5.3 The fall of the weight will be 5 ft *plus* 8 in. when it has compressed the spring 8 in. At that time the spring will have absorbed energy equal to $\frac{1}{2}ks^2 = \frac{1}{2} \times 20 \times 8^2 = 640$ lb-in., or 53.3 lb-ft.

The change in potential energy is equal to

$$\Delta \text{PE} = W \times h = 25(5 + 0.67) = 142 \text{ lb-ft}$$

a. Part of this energy has been absorbed by the spring, the remainder, 142 − 53.3, or 88.7 lb-ft, must remain as kinetic energy of the mass equal to $\frac{1}{2} Mv^2$. The velocity of the mass after it has compressed the spring is

$$v = \sqrt{\frac{88.7 \times 2}{M}} = \sqrt{\frac{88.7 \times 2g}{25}} = 15.1 \text{ fps}.$$

b. When the spring has been fully compressed, the velocity of the mass will become zero and all the change in potential energy of the mass will have been converted into potential energy of the spring.

$$W(5 \times 12 + s) = \frac{1}{2} \times 20 \times s^2 = 60 \times 25 + 25s = 10s^2$$

Rearranging and dividing by 5 throughout, we obtain

$$2s^2 - 5s - 300 = 0$$

By means of the binomial theorem,

$$s = \frac{5 \pm \sqrt{25 + 2400}}{4} = \frac{5 \pm 49.3}{4} = 13.6 \text{ in.}$$

5.4 The frequency of the fan assembly is

$$f = \frac{1}{2\pi} \sqrt{\frac{(I_f + I_m)k}{I_f I_m}}.$$

For the fan,

$$I_f = (10 \text{ lb/32.2}) \left(\frac{1}{12}\right)(9^2) = 2.1. \quad \text{(moment of inertia)}$$

For the motor,

$$I_m = (40 \text{ lb/32.2}) \left(\frac{1}{12}\right)(5^2) = 2.68. \quad \text{(moment of inertia)}$$

Also

$$k = 11.5 \times 10^6/9 \times 0.495 = 6.3 \times 10^5 \quad \text{(torsional constant)}$$

$$f = \frac{1}{2\pi} \sqrt{\frac{(2.1 + 2.68)(6.3 \times 10^5)}{2.1 \times 2.68}} = 116 \text{ cps.}$$

Motor frequency is 2400/60 = 40 cps. We may get excessive vibration when the system starts or is shut down if the rate of increase (or decrease) in speed is small, say around 40 cps.

In order to design the system without danger of excessive vibration, we should increase radius of gyration of both fan and motor armature and increase shaft diameter or decrease its length until frequency of the entire system is below 40 cps.

5.5 Let the following be understood:

$$n = 22 \text{ coils}$$
$$r = 3 \text{ in.}$$
$$d = 0.75 \text{ in.}$$
$$k = 1.18$$
$$G = 12 \times 10^6 \text{ psi}$$
$$\bar{i} = 85,000 \text{ psi}$$

a. The force to which the spring is stressed when loaded is calculated to be

$$P = 0.1963 \frac{0.75^3 \times 34,100}{3 \times 1.18} = 800 \text{ lb}$$

$$10 \text{ lb wt} \times \text{distance fired} = \frac{P \times \text{deflection}}{2}$$

$$10 \times s = (800 \times 8)/2 \quad \text{and} \quad s = \frac{800 \times 8}{2 \times 10} = 320 \text{ in.}$$

height to which projectile will be fired.
The actual stress is found to be

$$\frac{8 \text{ in.} \times 12 \times 10^6 \times 0.75 \text{ in.} \times 1.18}{4 \times 22 \times \pi \times 3 \text{ in.}^2} = 34,100 \text{ psi.}$$

b. The factor of safety is

$$\frac{\text{Stress at elastic limit}}{\text{Actual stress}} = \frac{85,000}{34,100} = 2.5.$$

5.6 With slider in position C the force pulling is F_1. This is equal to

$$F_1 = (14.1 - 2.5)12 = 139 \text{ lb.}$$

The horizontal and vertical components of the force are F_{1x} and F_{1y}, respectively.

$$F_{1x} = 139 \times 0.707 = 98.3 \text{ lb} = F_{1y}$$

With slider in position B as required,

$$F_2 = (7.65 - 2.5)12 = 62 \text{ lb}$$

where 7.65 is chord distance, in feet.

$$F_{2x} = 62 \times (7.07/7.65) = 57.3 \text{ lb}$$
$$F_{2y} = 62 \times (2.93/7.65) = 23.8 \text{ lb}$$

Then the work done is

$$\{[(98.3 + 57.3)2.93] + [(98.3 + 23.8)7.07]\}\frac{1}{2} = 660 \text{ in.-lb.}$$

5.7   The natural frequency of free vibration of the system in question is given by

$$f = \frac{1}{2\pi}\sqrt{\frac{kg}{W}}$$

where

k = spring modulus, lb per in.
g = gravitational constant, in. per $\sec^2$
W = weight on each spring, lb.

The amplitude of the forced vibration is given by A in the following equation:

$$A = x_s \frac{1}{1 - (f_1/f)^2}$$

where

x_s = displacement of spring caused by steady-state vibration
$f_1 = W/2\pi$, or frequency of exciting force, cps
f = natural frequency of system, cps

The ratio f_1/f is called the frequency ratio, and the ratio $1/[1 - (f_1/f)^2]$ may be interpreted as the *magnification factor*. Now, because of proportionality factors, we can say

$$\frac{20}{1200} = \frac{1}{1 - \left[\dfrac{(2000/60)^2}{f^2}\right]}.$$

For a reduction in the disturbing force, which we are after, the ratio A/x_s is negative. Now solve for f and this is found to be equal to 4.28 cps. Finally, substitute f in the first equation for natural frequency and

$$4.28 = \frac{1}{2\pi}\sqrt{\frac{k \times 386}{4000/6}}$$

and k is found to be equal to 1240 lb per in.

5.8   The frequency of the free vibration is f in accordance with

$$f = \frac{1}{2\pi}\sqrt{\frac{g}{6}} = \frac{1}{2\pi}\sqrt{\frac{386}{6}} = 1.28 \text{ cps.}$$

Because x_s is 1 in. and f_1 is 1 cps, the amplitude A of the forced vibration is

$$A = x_s \frac{1}{1-(f_1/f)^2} = 1 \times \frac{1}{1-(1/1.28)}2 = 2.56 \text{ in.}$$

Because the frequency of the impressed motion is below the resonant frequency, the motion of the axles is in phase with the motion of the body of the car. Hence, the change in the length of the spring is 2.56 – 1, or 1.56 in. The maximum shortening of the spring, therefore, is 1.56 + 6 = 7.56 in.

5.9   The momentum of the gun and carriage backward must be equal to the momentum of the projectile forward. Thus,

$$M_p v_p \cos\theta + M_g v_g = 0$$

where subscripts p and g are for projectile and gun (plus carriage).
Velocity of gun (and carriage) is found by rearranging the equation and

$$v_g = \frac{60 \times 1500 \times 0.8660}{3000} = 26 \text{ fps in a backward direction.}$$

5.10   The resulting stress is the sum of the normal running stress and that due to torsional impact because of the steel rod. The shearing stress during normal operation is

$$f_s = \frac{\text{Torque} \times 16}{\pi d^3}.$$

The torque must first be determined from the equation

$$\frac{\text{Hp} \times 33{,}000}{2\pi N} \times 12$$

Then
$$T = \frac{5 \times 33{,}000 \times 12}{2\pi \times 120} = 2650 \text{ in.-lb}$$

$$f_s = \frac{2650 \times 16}{\pi 1^3} = 13{,}500$$

On impact, all the kinetic energy of the rotating mass is instantly converted into potential energy, which would tend to disrupt the shaft. We may neglect the kinetic energy of the shaft and the mass moment of inertia of the impeller to simplify the calculations. Now, the kinetic energy of the motor is

$$E_m = 1/2 \times Iw^2 = 1/2 \times 10 \times (4\pi)^2 = 788 \text{ in.-lb}$$

where the angular velocity is w equal to $2\pi \times (\frac{120}{60})$, or 4π radians per sec. And I is equal to the mass moment of inertia of the motor as 10 in.-lb-sec^2. The shearing stress on the shaft is found by use of

$$S_s = \sqrt{\frac{4E_s E_k}{V_{shaft}}} = \sqrt{\frac{4(12 \times 10^6)788}{9.45 \text{ cu in.}}} = 63,200 \text{ psi.}$$

Finally, the maximum shear stress in the 1-in. shaft is

$$13,500 + 63,200 = 76,700 \text{ psi.}$$

5.11 Energy supplied by the flywheel is 4000 ft-lb total minus 800 ft-lb supplied by belt, leaving 3200 ft-lb to be supplied by flywheel. The energy given up by the flywheel is dependent on its change in rpm.

$$3200 = \frac{W}{2g} \times \rho^2 \times \left(\omega_1^2 - \omega_2^2\right)$$

Angular velocity $\omega_1 = 200 \times 2\pi/60 = 200 \times 2\pi/60 = 21$ radians per sec. Angular velocity ω_2 is similarly found to be equal to 18.9 radians per sec. Then substituting in the equation

$$3200 = \frac{W}{64.4} \times \left(\frac{18}{12}\right)^2 (21^2 - 18.9^2)$$

from which W is found to be equal to 1090 lb.

5.12 Speed of rotation of a shaft and the rotor attached to it at which resonance occurs is frequently called the critical speed for the shaft. In general, it is important to know the critical speed of a rotating member so that speed of operation can be maintained either considerably above or below this dangerous (critical) speed, or so that the member can be designed for an operating speed that will not be too near to the critical speed. To calculate the critical speed, first determine equivalent spring modulus of the shaft with a concentrated load W (fan wheel) at the center. Spring modulus $k = W/\Delta$, for which Δ is static deflection for a beam with concentrated load *at center*,

$$\Delta = \frac{Wl^3}{48EI} \text{ in.}$$

where l is distance between bearings, in.; E is modulus of elasticity (for steel it's 30×10^6 psi); and I is moment of inertia for the circular cross-sectional area of diameter d, that is, $\pi d^4/64$. Spring modulus may be found to be equal to $k = 48EI/l^3$, lb per in. If we neglect weight of shaft, the natural frequency or critical speed is

$$S_c = \frac{60}{2\pi}\sqrt{\frac{kg}{W}} = \frac{60}{2\pi}\sqrt{\frac{48EI}{l^2} \times \frac{386}{W}}.$$

a. By decreasing span length 10 percent, with all other items constant, we can apply a correction factor to the original S_{c1}. Thus,

$$S_{c2} = S_{c1}\sqrt{\frac{1}{l^3}} = S_{c1}\sqrt{\frac{1}{(1-0.1)^3}} = S_{c1} \times 1.16.$$

Critical speed will be increased 16 percent.

b. By increasing shaft diameter 10 percent

$$S_{c2} = S_{c1}\sqrt{d^4} = S_{c1}\sqrt{1.1^4} = S_{c1} \times 1.22.$$

Critical speed will be increased 22 percent.

c. By increasing rotor weight 10 percent

$$S_{c2} = S_{c1}\sqrt{\frac{1}{W}} = S_{c1}\sqrt{\frac{1}{1.1}} = S_{c1} \times 0.953$$

Critical speed will be decreased 4.7 percent.

5.13 Each spring carries 25 lb, assuming uniform distribution.

$$f = \frac{1}{2\pi}\sqrt{\frac{kg}{W}} = \frac{1}{2\pi}\sqrt{\frac{15 \times 386 \text{ in. per sec}^2}{25}} = \frac{15}{6.28} = 2.4 \text{ cps}$$

The period is $1/f = 1/2.4 = 0.417$ sec.

5.14 This is an example of an evenly distributed weight that vibrates because of the elasticity of the shaft rather than that of a spring. To solve for the frequency f, which equals $(1/2\pi)\sqrt{g/\Delta}$, it is necessary to determine the static deflection Δ of the system. We know that a simply supported shaft (or beam) has $\Delta = (5Wl^3)/(386EI)$, where E is the tensile modulus of elasticity and I is the moment of inertia of the cross-sectional area about the neutral axis. Finally

$$f = \frac{1}{2\pi}\sqrt{\frac{386 \times E \times I}{5Wl^3}} \text{ cps}$$

Note 386 is gravitational constant g, in. per sec^2. Thus, for similar systems merely substitute for Δ the static deflection from tables or calculated values.

5.15 Refer to Exhibit 5.15. Rotor revolves with angular velocity ω (radians per sec) and unbalanced mass M at a distance r (ft) from the axis of rotation, causing a rotating unbalance (centrifugal) force equal to $Mr^2\omega = P_0$. In a forced vibration, if the period of the impressed force is the same as that of the free or natural period of vibration of the system, the theoretical amplitude of the vibration becomes exceedingly large. This condition is known as *resonance* and is, of course, to be avoided in parts of machines and structures, i.e., natural frequency equals resonant frequency, $f_n = f_r$. Thus

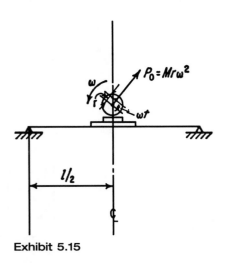

Exhibit 5.15

$$f_r = f_n = \frac{1}{2\pi}\sqrt{\frac{kg}{W}} = \frac{1}{2\pi}\sqrt{\frac{g}{\Delta}} = \frac{1}{2\pi}\sqrt{\frac{386}{0.5}} = 4.42 \text{ cps}$$

a. Rotating speed at resonance $= f_r \times 60 = 265.2$ rpm.
b. The amplitude of the forced vibration is the product of the static deflection caused by the rotating unbalanced force and the *amplification factor*

$$\frac{P_0}{k}\left[\frac{1}{1-(f_1/f_n)^2}\right] \text{ in.}$$

where f_1/f_n is frequency ratio. Continuing,

$$P_0 = \frac{W}{g}r\omega^2 = \frac{2}{32.2} \times 1 \times \left(\frac{2\pi \text{ rmp}}{60}\right)^2 = 89 \text{ lb}$$

Then
$$f_1 = \frac{\omega}{2\pi} = \frac{12\pi}{2\pi} = 6 \text{ cps}$$

Also
$$f_1 = (\text{rpm}/60) = \frac{360}{60} = 6 \text{ cps}$$

Finally

$$\frac{89}{W/\Delta}\left[\frac{1}{1-(6/4.42)^2}\right]=\frac{89}{(25/0.5)}\left[\frac{1}{1-1.85}\right]=1.78(-1.178)=-2.1 \text{ in.}$$

The value of the amplification factor can be positive (+) or negative (−) depending on whether $f_1 < f_n$ or not. At resonance ($f_1 = f_n$) the amplitude is theoretically infinite. Actually, however, the damping, which is always present, holds the amplitude to a finite amount.

5.16 Refer to Exhibits 5.16a and 5.16b.

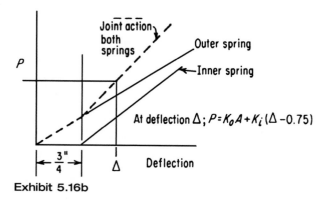

Exhibit 5.16b

In general, $K = d^4G/8D^3N$ for spring constant. G for carbon steel is 11.5×10^6 psi. Therefore,

$$K_0 = \frac{(1.5)^4 \times (11.5 \times 10^6)}{8 \times 7.5^3 \times 6} = 2870 \text{ lb per in.}$$

$$K_i = \frac{(1^4) \times (11.5 \times 10^6)}{8 \times 4.5^3 \times 9} = 1750 \text{ lb per in.}$$

Because

$P = K_0\Delta + K_i(\Delta - 0.75)$, then $20,000 = 2870\Delta + 1750(\Delta - 0.75)$
$\Delta = 21,310/4620 = 4.6$ in.

Outer spring $\Delta_0 = 4.6$. Inner spring $\Delta_i = 4.6 - 0.75 = 3.85$

Therefore,

$$P_0 = K_0\Delta_0 = 13,200 \text{ lb } P_i = K_iA_i = 6720 \text{ lb}$$

5.17
$$T = \frac{63,000 \times \text{hp}}{N} = \frac{63,000 \times 25}{900} = 1750 \text{ in.-lb}$$

$$Ks = \frac{6M}{bh^2} \frac{3c_1^2 - c_1 - 0.8}{3c_1(c_1 - 1)}$$

$$30,000 = \frac{6 \times 1750}{0.75^3} \frac{3c_1^2 - c_1 - 0.8}{3c_1(c_1 - 1)}$$

$$c_1^2 - 4.25c_1 + 1.302 = 0 \quad c_1 = 3.922$$

$$R = \frac{c_1 d}{2} = \frac{3.922 \times 0.75}{2} = 1.47 \text{ in.}$$

$$\phi = \frac{Ml}{EI} \quad l = \frac{\phi EI}{M} = \frac{\phi Ebh^3}{12M} = \frac{5\pi}{180} \times \frac{30 \times 10^6 \times 0.75^4}{12 \times 1750}$$

$$= 39.4 \text{ in.}$$

$$N = \frac{l}{2\pi R} = \frac{39.4}{2\pi 1.47} = 4.27 \text{ coils}$$

where

c_1 = spring index
d = diameter of wire
E = modulus of elasticity
I = moment of inertia
K = spring rate
N = number of active coils
R = mean radius of helix
T = torque

Pumps and Hydraulics

PROBLEMS

6.1 A centrifugal pump delivers 300 gpm of water at 3000 ft total dynamic head when operating at 3500 rpm.
   a. At what speed must a geometrically similar pump operate to deliver 200 gpm at the same total dynamic head?
   b. What must be the diameter of the impeller of this new pump if the diameter of the 3500-rpm pump is 6 in.?

6.2 A boiler-feed pump is driven by a variable–speed steam turbine. The characteristic curve of the pump at a speed of 6000 rpm and the system-head curve of the boiler–feed-water system are shown in Exhibit 6.2. Determine the speed at which the pump should be operated to match a system requirement of 3000 gpm.

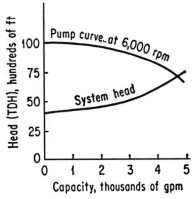

Exhibit 6.2

6.3 A model centrifugal pump with a 3-in.-diameter impeller delivers 600 gpm of 60°F water at a total head of 350 ft when operating at 1750 rpm. Find the diameter of a geometrically similar pump that will deliver 1000 gpm when operating at 3500 rpm. What will be the total head of the 3500-rpm pump when it is delivering 1000 gpm?

6.4   A dc-motor-driven pump running at 100 rpm delivers 500 gpm of water against a total pumping head of 90 ft with a pump efficiency of 60 percent.
  a.   What motor horsepower is required?
  b.   What speed and capacity would result if the pump rpm were increased to produce a pumping head of 120 ft, assuming no change in pump efficiency?
  c.   Can a 25-hp motor be used under conditions indicated in (b)?

6.5   A client purchased and installed a new centrifugal pump for which the manufacturer estimated the following performance based on design calculations: speed 250 rpm, total dynamic head 25 ft of water, efficiency 84.5 percent, brake horsepower of driving motor 750, all at rated capacity of 100,000 gpm. A test yielded the information shown in Exhibit 6.5. All data shown were rechecked for accuracy and found correct. Flow rate could not be measured due to size of pump. On the basis of the above, what would you tell your client? For a more complete answer, what additional test data would you require? How would you propose to obtain it?

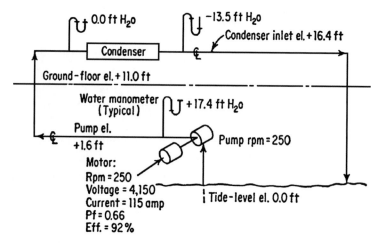

All piping smooth clean concrete conduit, 78 in. ID
Equivalent length of conduit, pump to condenser inlet, 1.84 ft
Calculated friction drop, pump to condenser inlet, 2.6 ft H₂0
Calculated friction drop across condenser, 13.3 ft H₂0
Calculated friction drop, condenser outlet to river 1.3 ft H₂0

Exhibit 6.5

6.6   A fire breaks out in a three-story warehouse. A pressure pump engine has a $2\frac{1}{2}$-in. fire hose, 300 ft long, with a nozzle 50 ft above the pump. If the discharge is 250 gpm and $C = 100$ (William and Hazen) or $n = 0.009$ (Kutter and Manning), the pressure behind the nozzle is 28 psi and minor losses are neglected.
  a.   Compute pressure at the pump.
  b.   Draw the hydraulic gradient line.
  c.   If the system is 80 percent efficient and there is a pressure of 10 psi at the pump inlet, find the horsepower required.

6.7 Vacuum pumps are often used at altitudes high enough (Denver or Mexico City, for example) to appreciably affect their performance characteristics. This altitude factor is often overlooked by design and plant engineers, but if ignored, it can spell trouble for air-operated equipment.

    a. A vacuum pump's capability can be measured in terms of the percentage of atmosphere it can exhaust from a closed system. If a barometer at sea level registers air pressure of 30 in. Hg under standard conditions, what would be the maximum vacuum rating of a vacuum pump that has a maximum capability of 24 in. Hg?

    b. What will a vacuum pump rate at 5000 ft altitude in airflow production and maximum vacuum Hg if at sea level it is rated to provide a flow of 10 cfm with 25 in. Hg maximum suction?

6.8 At the higher end of its capacity range, a centrifugal pump is conveniently self-cooling; the volume of fluid moving through the pump is sufficient to cool off the pump by carrying away the heat generated by the pump's rotating parts and bearings. However, at lower flow rates, particularly when a pump is operating at less than 25 percent capacity, the volume of fluid being discharged can become inadequate for cooling, increasing the risk of overheating and seizure of bearings.

    a. To protect centrifugal pumps from overheating damage and instability under low-flow conditions, what protective measures must be instituted?

    b. Discuss each of these measures with advantages and disadvantages.

    c. Select one of the systems and sketch the pump, piping, and valving.

SOLUTIONS

6.1   a. This is an application of the specific speed of the pumps. First determine the specific speed of the existing pump, and then use this work to obtain the speed sought.

$$N_s = \frac{N_1\sqrt{Q_1}}{H^{3/4}} = \frac{3500\sqrt{300}}{3000^{3/4}} = \frac{3500 \times 17.3}{420} = 144$$

For the new pump

$$144 = \frac{N_2\sqrt{200}}{420} \qquad N_2 = \frac{144 \times 420}{\sqrt{200}} = 4277 \text{ rpm.}$$

  b. Two geometrically similar centrifugal pumps will have similar flow conditions if the ratio of the fluid velocities to the velocities of the rotating parts is the same, i.e., if

$$\frac{Q_1}{N_1 D_1^3} = \frac{Q_2}{N_2 D_2^3}$$

Thus, for the problem at hand, insert the proper values in the preceding relation.

$$\frac{300}{3500 \times 6^3} = \frac{200}{4250 \times D_2^3}$$

$$D_2^3 = \frac{200 \times 3500 \times 6^3}{4250 \times 300} = 118.5 \quad D_2 = 5 \text{ in. even}$$

Note that for two geometrically similar pumps the following conditions also exist, where Q = cfs, H = head ft, P = power, N = rpm, D = impeller diameter, in.

$$\frac{H_1}{\left(Q_1/Q_1^2\right)^2} = \frac{H_2}{\left(Q_2/D_2^2\right)^2} \quad \frac{H_1}{(N_1 D_1)^2} = \frac{H_2}{(N_2 D_2)^2}$$

And for the same fluid

$$\frac{P_1}{Q_1 H_1} = \frac{P_2}{Q_2 H_2} \quad \frac{P_1}{N_1^3 D_1^5} = \frac{P_2}{N_2^3 D_2^5}$$

6.2   Refer to Exhibit 6.2. Because capacity is directly proportional to speed and the system curve holds, select a gpm at 6000 rpm and set up a proportion.

At 3000 gpm and 6000 rpm, head is 92 ft.
At 3000 gpm and x rpm, head is 50 ft.

$$\frac{92}{50} = \left(\frac{6000}{x}\right)^2 = 1.84 = 36 \times \frac{10^6}{x^2} \quad x^2 = \frac{36 \times 10^6}{1.84} = 19.6 \times 10^6$$

$$x = 4320 \text{ rpm.}$$

6.3   Review Problem 6.1. Note the additional equations.

Diameter:

$$D_2 = \frac{Q_2 N_1 D_1^3}{Q_1 N_2} = \frac{1000 \times 1750 \times 3^3}{600 \times 3500} = 2.25 \text{ in.}$$

Total head:

$$H_2 = H_1 \frac{N_2^2 D_2^2}{N_1^2 D_1^2} = 350 \times \frac{3500^2 \times 2.25^2}{1750^2 \times 3^2} = 785 \text{ ft}$$

6.4   a.
$$\frac{500 \times 8.33 \times 90}{33,000 \times 0.60} = 18.9$$

Use 20-hp motor.

b.   By application of the laws of affinity

$$\frac{90}{120}=\frac{100^2}{x^2} \quad x^2=100^2\times\frac{120}{90}=1.33\times100^2 \quad x=1.151\times100.$$

The new rpm will be 11.5. The new capacity will be

$$\frac{500}{y}=\frac{100}{115} \quad y=500\times\frac{115}{100}=575 \text{ gpm}.$$

c.   Again by application of the laws of affinity

$$\frac{18.9}{z}=\frac{100^3}{115^3} \quad z=18.9\times115^3/100^3=18.9\times1.52=28.7 \text{ bhp}$$

A 25-hp motor cannot be used. Use a 30-hp motor.

6.5   A quick check of pipe friction shows this to be in order. This was done using the Darcy equation with a friction factor of 0.02 for clean, new steel pipe. Next check TDH of 25 ft, using Bernoulli's equation. Velocity and pressure heads cancel out, from which TDH is found to be 17.2 ft. This figure would actually be greater because of suction pipe loss, although it would be too small to be included. Now check brake horsepower from calculated figures.

$$\text{Bhp}=\frac{\text{gpm}\times8.33\times17.2}{33,000\times0.845}=\frac{100,000\times8.33\times17.2}{33,000\times0.845}=515$$

Electrical horsepower is now checked, assuming 3-phase ac power.

$$\frac{1.73\times I\times E\times\text{eff}\times\cos\Phi}{746}=\frac{1.73\times115\times4150\times0.92\times0.66}{746}$$

Result is 673 hp. The motor horsepower was selected originally on the basis of

$$\text{Bhp}=\frac{100,000\times8.33\times25}{33,000\times0.845}=746. \text{ This is acceptable.}$$

Although all pressure has been consumed to the point of pipe entry to the condenser, the losses through condenser and in the line back to tidewater are more than compensated for by the hydrostatic leg from elevation plus 16.4 to elevation 0.00 ft. I would tell the client that the system is in order. However, there are further checks that would help confirm these findings. Condenser operation should be checked for vacuum so that water flow is as required. Pump curve and system curve superimposed thereon would give the final check for performance point. If this lines up with the system curve, this clinches it as performing to expectations.

6.6   a.   Use Bernoulli's theorem between *A* and *B* shown in Exhibit 6.6 with the line through *B* as the datum.

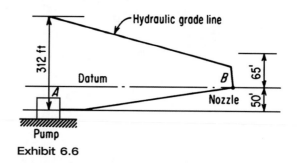

Exhibit 6.6

$$\frac{V_A^2}{2g}+\frac{P_A}{W}+Z_A=\frac{V_B^2}{2g}+\frac{P_B}{W}+Z_B+H_f$$

Velocity heads are equal and thus cancel out.

$$\frac{P_A}{W}+Z_A=\frac{P_B}{W}+Z_B+H_f$$

$$\frac{P_A}{W}+0=\frac{28}{0.433}+50+H_f$$

$$P_A=62.4(65+50+197)=62.4(312)\ \text{psf}$$

$$=62.4\left(\frac{312}{144}\right)=135\ \text{psi at the pump}$$

b.   Hydraulic gradient: see Exhibit 6.6.

c.

$$\text{Hp of pump}=\frac{\text{lb water per minute}\times\text{head}}{33,000\times\text{Efficiency}}$$

$$=\frac{8.34\times250\times(135-10)\times2.307}{33,000\times0.80}$$

$$=22.8\ \text{brake hp}$$

Use a 25-hp motor.

6.7   a.   Maximum vacuum rating $=\frac{24}{30}=0.8$ or 80 percent. This is the percentage of atmosphere it can exhaust from a closed system.

b.   A 10-cfm vacuum pump with a 25-in. Hg maximum rating at sea level will produce an airflow of only $10\times0.8=8$ cfm at 500 ft altitude and will result in $25\times0.8=20$ in. Hg vacuum. This pump would not provide sufficient capacity. A model with a sea-level airflow of 12 cfm would be required, since $12\times0.83=10$ cfm.

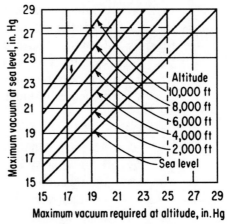

Exhibit 6.7a

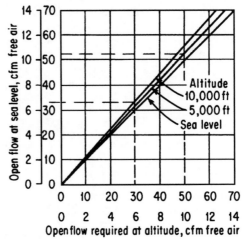

Exhibit 6.7b

Maximum vacuum possible decreases with increasing altitude. Exhibit 6.7a graphically illustrates this relationship. Exhibit 6.7b illustrates altitude vs. open flow requirements. For Exhibit 6.7b, the relationship between altitude and maximum vacuum may be determined.

6.8 a. Recirculation systems, also known as bypass systems, are necessary, and may take the form of continuous-recirculation systems, flow-controlled recirculation systems, and self-regulated recirculation systems.

b. *Continuous recirculation:* Because of increasing electrical rates, continuous-recirculation systems, which recirculate fluid regardless of whether it is needed to protect the pump, are becoming cost-prohibitive. At 3 cents per kWh, for example, the electrical power costs for recirculating 700 gal/min at a discharge head of 500 ft are $27,000 annually. In addition, because this system bypasses fluid even when process demand is at a maximum, capital costs are increased by the need to oversize the pump and its prime mover at the outset. Check the above operating costs roughly, using water as the fluid, by the following formula:

$$\frac{700 \times 8.33 \times 500 \times 8640 \text{ h/year} \times 746 \times 0.03}{33,000 \times 0.6} = \$28,472$$

Flow-controlled recirculation: These systems reduce electrical consumption by recirculating fluid only when flow approaches a minimum safe rate. However, while energy and pump size are reduced, these systems are both complex and costly. A multiplicity of components increases the problem of breakdowns, and the need for instrument loops increases installation, operating, and maintenance costs.

Self-regulated recirculation: These systems are finding wider use and acceptance in both utility and process applications. They offer economies in the areas of operation, installation, and maintenance and combine all the functions of flow sensing, reverse-flow protection, pressure breakdown, and on-off or modulated bypass flow in a single unit. Capable of handling high recirculation pressure drops, self-powered units employ a rising-disk type of check valve that acts as the flow-sensing element. The system requires only three connections.

c.   Refer to Exhibits 6.8a–c.

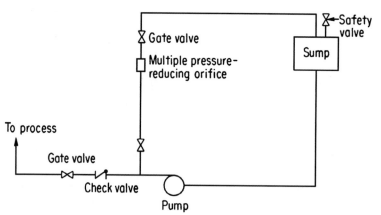

Exhibit 6.8a Continuous recirculation system

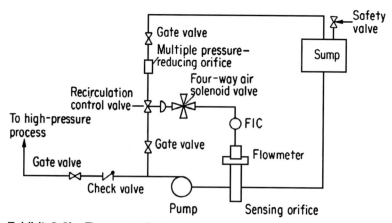

Exhibit 6.8b Flow-controlled recirculation system

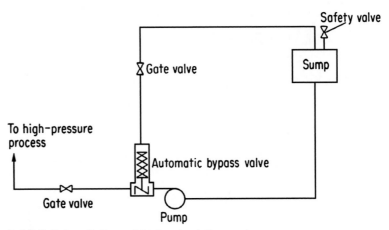

Exhibit 6.8c Self-regulated recirculation system

Fans, Blowers, and Compressors

OUTLINE

PROBLEMS

7.1   The characteristic and horsepower curves shown in Exhibit 7.1a indicate the performance of a double-inlet forced-draft fan operating at sea level with a speed of 1180 rpm when handling air at 40°F. Determine the static pressure developed by the fan and the power required to drive it when it handles 514,000 lb per hr of 80°F air at an altitude of 5000 ft, where the barometric pressure is approximately 24.9 in. Hg abs.

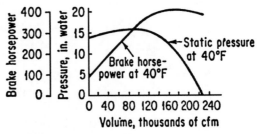

Exhibit 7.1a

7.2   a.   Draw an approximate head vs. capacity curve for a fan handling flue gas of any standard type or a design pressure of 10 in. water and 150,000 cfm at 300°F.

   b.   Assuming the fan to be driven by a constant-speed motor, explain how to determine the head curve with gas at 400°F.

   c.   If a 30 percent pressure drop is added to the system resistance, what will be the output of the fan in cfm with the gas at 300°F?

7.3 Characteristic curves in Exhibit 7.3 are for a forced-draft fan operated at speeds of 1200 and 900 rpm. Also shown in the efficiency curve obtained at an operating speed of 1200 rpm. Estimate the horsepower required to drive the fan when operated at a speed of 900 rpm and at a flow rate of 300,000 cfm at 15 in. water static pressure.

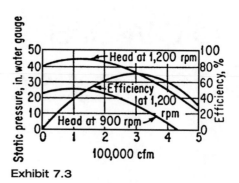

Exhibit 7.3

7.4 The performance characteristics of a centrifugal compressor are shown in Exhibit 7.4; symbols are defined as follows:

P_1 = compressor inlet pressure, absolute
P_2 = compressor discharge pressure, absolute
T_1 = compressor inlet temperature, °R
M = mass flow rate, lb per sec
N = compressor speed, rps

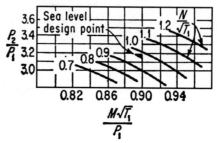

Exhibit 7.4

The compressor is designed to operate at sea level at 14.7 psia and 60°F. Estimate in psia the compressor exit pressure if the compressor is operated at the same speed and at the same volume flow rate in Denver, Colorado, on a day when the atmospheric pressure and temperature are 12 psia and 0°F, respectively.

7.5 The delivery pressure of a reciprocating compressor having isentropic compression and handling 0.1 lb of air per cycle is 50 psia. The temperature of the air drawn in is 66°F and the average atmospheric pressure where the compressor is located may be taken as 14.7 psia. Determine (a) the net work of the compressor cycle if the compressor has no clearance; (b) the net work of the compressor cycle if the compressor has 5 percent clearance; (c) the weight of the air in the cylinder during the compression if the compressor has 5 percent clearance; (d) the percent increase in the required piston displacement with 5 percent clearance as compared to a compressor of the same capacity with no clearance.

7.6 Assume three compressor cylinders without clearance and with a piston displacement of 1 cu ft, one cylinder so arranged as to give adiabatic compression, one arranged to give isothermal compression, and one arranged to give a compression curve with an exponent equal to 1.25. (a) Determine the work during the compression process alone in each cylinder and the final temperature in each case if air with an initial temperature of 60°F and an initial pressure of 14.7 psia is compressed to 100 psia. (b) Express the second and third cases as a percentage of the compression work in the case of the adiabatic process. (c) Determine the work per cycle in each case, assuming discharge to occur at the constant pressure of 100 psia. (d) Express results as percentages as in (b).

7.7 What is the volumetric efficiency of a compressor having 5 percent clearance and operating as in question 7.5?

7.8 A centrifugal compressor handling air draws in 12,000 cu ft of air per minute at a pressure of 14 psia and a temperature of 60°F. This air is delivered from the compressor at a pressure of 70 psia and a temperature of 164°F. The area of the suction pipe is 2.1 sq ft, the area of the discharge pipe is 0.4 sq ft, and the discharge pipe is located 20 ft above the suction pipe. The weight of the jacket water, which enters at 60°F and leaves at 110°F, is 677 lb per min. Find the horsepower required to drive this compressor assuming no loss from radiation.

7.9 Air is admitted to the cylinder of a 12- by 16-in. air engine under pressure of 100 psia until one-fourth of the stroke is reached and then is allowed to expand to the end of the stroke following the law $PV = C$. Find the work done between cutoff and end of stroke.

7.10 An ammonia oxidation unit will use 16,000 cfm of air (60°F and 1 atm), compressing it from standard conditions of pressure to 125 psig in a centrifugal compressor with nine stages and no intercoolers. Part of the power to drive the compressor will be supplied by a turboexpander, which is fed with the waste nitrogen from the nitric acid absorption towers at 75 psig. It may be assumed that this waste gas is pure nitrogen and that all of it was derived from the air. The nitrogen from the absorbers is preheated to 650°C by the effluent gases from the converter and enters the expander at this temperature. The remainder of the power required for air compression is to be supplied by an electric motor. Estimate this motor in horsepower.

7.11 A single-stage compressor is designed to compress air at standard conditions of pressure but at 80°F to a pressure of 80 psig. The clearance is estimated to be 8 percent. Estimate the percentage change in the capacity of the same compressor (expressed as percent) and in the horsepower needed to drive it if it were located at an elevation where the air pressure is only 0.5 atm absolute. Discharge gauge pressure remains the same. Estimate the increase in the maximum temperature of the air in the cylinder.

7.12 A centrifugal compressor driven by a 2000-hp steam turbine has been used to compress methane from 60°F and 14.7 psia to 150 psia. (a) It is desired to change over to air compression with the same intake conditions and discharge pressure. Will the compressor perform satisfactorily? (b) A need has arisen for 1500 lb of saturated steam per hr at 800 psia but the steam is available only at 200 psia. It is proposed to compress the given steam to the desired pressure in a continuous reciprocating compressor. Estimate the work required in kwhr per 100 lb of saturated steam at 800 psia.

7.13 A 5-stage centrifugal compressor running at 4330 rpm is tested with air at 120°F and standard atmospheric pressure. At a capacity of 16,000 cfm at inlet conditions, the discharge pressure was 32.5 psig and the brake hp was 2100. Calculate the following quantities: (a) adiabatic head; (b) adiabatic power efficiency; (c) polytropic exponent; (d) polytropic efficiency; (e) fluid or polytropic head; (f) pressure that would be developed if the compressor were fed carbon dioxide at 60°F and 3 psig; (g) final temperature of the carbon dioxide; (h) brake horsepower required when the feed is carbon dioxide. Note compressor has no intercoolers.

7.14 For a 40-in.-diameter fan delivering 160,000 cfm of standard air with 16 in. static pressure (SP) at 1200 rpm, what is the diameter of a geometrically similar fan running at 60,000 rpm that will deliver 1.0 cfm and 1 in. SP?

7.15 A compressor is used to supply 50 lb/in.^{2} gauge air to an air-lift pump that is raising 20 gal/min of a liquid with specific gravity of 1.5 to a height of 50 ft. Pump efficiency is 30 percent, and it may be assumed that the compressor operates isentropically ($C_p/C_v = 1.4$). Determine the horsepower required for the compressor.

SOLUTIONS

7.1 The fan will handle the same cfm no matter where located as long as the rpm and fan configuration does not change. However, because air in the actual location is less dense than the curve characteristic indicates, both static pressure and bhp developed will be less than at sea level. The cfm handled under all conditions may be obtained in the usual manner.

$$\frac{514,000}{60} \times \frac{379}{29} \times \frac{460+80}{460+60} \times \frac{29.9}{24.9} = 139,550 \text{ cfm}$$

At sea level, from the characteristic curve at 40°F, the static pressure reads 16 in. WG. The brake horsepower reads to be equal to 325. Because brake horsepower and static pressure will vary directly with relative density, use Exhibit 7.1b to find this factor. Correction factor reads 0.90, the relative density factor.

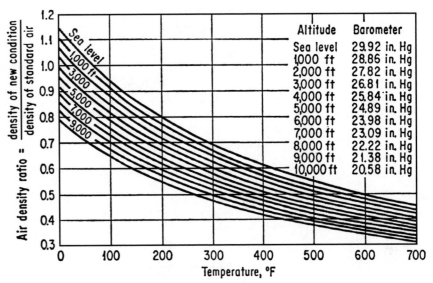

Exhibit 7.1b

New static pressure = $16 \times 0.90 = 14.4$ in. WG

New bhp = $325 \times 0.90 = 292.5$ bhp

7.2 a. See Exhibit 7.2. Fan manufacturer will guarantee design point only. When purchasing a fan (or pump), the fan characteristic curve should be made a part of the specification.

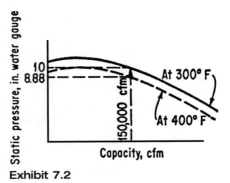

Exhibit 7.2

b. As indicated in Problem 7.1, the cfm will always remain practically the same so long as the rpm does not change. However, the static pressure will decrease with a decrease in density, and with increase in temperature (from 300 to 400°F) the air density ratio will become less than unity. Assume fan at sea level, and refer to Exhibit 7.1b. The density ratio is read off to be 0.62 at 400°F and 0.70 at 300°F. The correction factor to be applied to the 10 in. is 0.62/0.70 = 0.888. The new static pressure for the 400°F condition is 8.88 in. WG. Locate this point along the 150,000 cfm line where it crosses the 8.88-in. line from the left side of the figure. This locates the locus of points and runs parallel to the 300°F curve and establishes the new head vs. capacity curve.

c.   Assume constant speed. Then there will be a reduction in flow in accordance with

$$\text{New cfm at } 300°F = 150,000 \times \sqrt{\frac{1}{1+0.3}}$$
$$150,000 \times 0.877 = 131,550 \text{ cfm.}$$

7.3   Refer to Exhibit 7.3. At 900 rpm read curve left for 15 in. static pressure and 300,000 cfm. At 1200 rpm the new static pressure developed is

$$\frac{15}{y} = \frac{1200^2}{900^2} \qquad y = 15 \times \frac{1200^2}{900^2} = 26.7 \text{ in. WG}$$

The new cfm is

$$\frac{300,000}{x} = \frac{900}{1200} \qquad x = 300,000 \times \frac{1200}{900} = 400,000 \text{ cfm.}$$

From Exhibit 7.3 for 400,000 cfm and 1200 rpm the fan efficiency is read to be equal to 62 percent. Assuming the laws of affinity apply and efficiency remains constant at 62 percent, the horsepower required is by use of a recognized formula,

$$\text{bhp} = 0.000160 \times 300,000 \times 15 \times 1/0.62 = 1160.$$

7.4   For sea-level compressor operation the following apply:

$$P_1 = 14.7 \text{ psia} \quad \text{from curve,} \quad \frac{M\sqrt{T_1}}{P_1} = 0.9 \quad \frac{P_2}{P_1} = 3.2$$

$$T_1 = 520°R \quad \text{from curve,} \quad \frac{N}{\sqrt{T_1}} = 1$$

The mass flow rate at sea level is first calculated, converted to cfm. Because the compressor is a constant-volume machine, the new weight rate can be determined for Denver conditions.

$$M = 0.9 \times 14.7/520^{1/2} = 30.2 \text{ lb per sec}$$

$$\text{cfs} = \frac{379}{29} \times 30.2 \times (14.7/14.7) \times \left(\frac{520}{520}\right) = 395 \text{ cfs}$$

The new M in Denver is equal to

$$395 \times \frac{29}{379} \times 12/14.7 \times \frac{520}{460} = 27.8 \text{ lb per sec}$$

Because we are asked to estimate the exit pressure, we can assume that the work in both locations is the same. Thus,

$$20.2 \times c_p \times 520[1 - 3.2^{0.26}] = 27.8 \times c_p \times 460 \left[1 - \left(\frac{P_2}{P_2} \right)^{0.26} \right]$$

where $0.26 = (n-1)/n = (1.35-1)/1.35$. Because the c_p's cancel out, it may be shown that $(P_2/P_1)^{0.26} = 1.43$. To find P_2/P_1, proceed as follows:

$$0.26 \log \frac{P_2}{P_1} = \log 1.43 \qquad \log \frac{P_2}{P_1} = \frac{0.1553}{0.26} = 0.598$$

$$\frac{P_2}{P_1} = 3.96 \qquad P_2 = 12 \times 3.96 = 47.5 \text{ psia.}$$

7.5   a.   Refer to Exhibit 7.5a (*PV* diagram) and Exhibit 7.5b (*TS* diagram) for no clearance.

$$\text{Net work} = \frac{k}{k-1} wRT_{2i} \left[\left(\frac{P_3}{P_2} \right)^{(k-1)/k} - 1 \right]$$

$$= \frac{1.4}{0.4} \times 0.1 \times 53.3 \times 520 \left[\left(\frac{50}{14.7} \right)^{0.286} - 1 \right]$$

$$= 4055 \text{ ft-lb.}$$

b.   The clearance does not affect the net work delivered by the compressor when handling a given volume of air or a given weight of air. Therefore, the net work for 5 percent clearance is 4055 ft-lb.

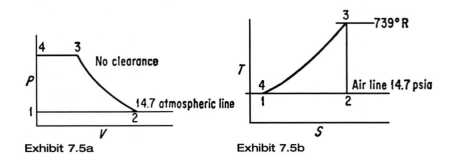

Exhibit 7.5a   Exhibit 7.5b

c.   Percent clearance is equal to

$$m = \frac{\text{clearance volume}}{\text{piston displacement}} = 0.05$$

$$V_1 = 1 + m - m\left(\frac{P_3}{P_2}\right)^{1/n} = 1 + 0.05 - 0.05\left(\frac{50}{14.7}\right)^{1/1.4} = 0.93$$

$$V_3 = 1.24/0.93 = 1.332 \qquad 1.332 \times 0.0807 = 0.113\,\text{lb}$$

d.
$$\frac{1.332 - 1.24}{1.24} \times 100 = 7.4\,\text{percent}$$

7.6   a.   Refer to Exhibit 7.6

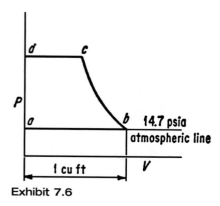

Exhibit 7.6

Adiabatic compression cylinder:

$$\frac{T_1}{T_2} = \left(\frac{P_1}{P_2}\right)^{(k-1)/k} = \left(\frac{14.7}{100}\right)^{0.286} = 0.578 \qquad T_2 = \frac{520}{0.578} = 900°\text{R}, \quad \text{or} \quad 440°\text{F}$$

$$\text{Work of compression} = \frac{P_3 V_3 - P_2 V_2}{k - 1}$$

$$\frac{V_3}{V_2} = \left(\frac{P_4}{P_3}\right)^{1/k} = 0.254 \qquad V_3 = 0.254 \times 1 = 0.254\,\text{cu ft}$$

$$\text{Work} = \frac{(100 \times 25.4) - (14.7 \times 1)}{1.4 - 1} \times 144 = \frac{10.7}{0.4} \times 144$$

$$= 3860\,\text{ft-lb}$$

Isothermal compression cylinder:

$$T_3 = T_2 = 520°\text{R}, \quad \text{or} \quad 60°\text{F}$$

$$\text{Work} = P_2 V_2 \ln\frac{V_2}{V_3} = 14.7 \times 144 \times 1 \times \ln\frac{1}{0.254}$$

$$= 14.7 \times 144 \times 1 \times 1.3712$$

$$= 2900\,\text{ft-lb}$$

Polytropic compression cylinder:

$$T_3 = T_2 \left(\frac{100}{14.7} \right)^{(k-1)/k} = 520 \times 1.468 = 763°R, \quad \text{or} \quad 303°F$$

$$\text{Work} = \frac{P_3 V_3 - P_2 V_2}{n-1} = \frac{(100 \times 144 \times 0.209) - 2120}{0.25}$$

$$= \frac{3010 - 2120}{0.25}$$

$$= 3560 \text{ ft-lb}$$

b. $\qquad \frac{2900}{3850} \times 100 = 75.5 \text{ percent} \qquad \frac{3560}{3850} \times 100 = 92.5 \text{ percent}$

c. $\qquad \text{Net work adiabatic} = \frac{k}{k-1} P_2 V_2 \left[\left(\frac{P_3}{P_2} \right)^{(k-1)/k} - 1 \right]$

$$= \frac{1.4}{0.4} \times 14.7 \times 144 \times 1 \left[\left(\frac{100}{14.7} \right)^{0.4/1.4} - 1 \right]$$

$$= 3.5 \times 14.7 \times 144 \times 1(1.728 - 1) = 5380 \text{ ft-lb}$$

$$\text{Net work isothermal} = P_2 V_2 \ln \frac{P_3}{P_2} = 14.7 \times 144 \times 1 \times 1.9184 = 4055 \text{ ft-lb}$$

$$\text{Net work polytropic} = \frac{n}{n-1} P_2 V_2 \left[\left(\frac{P_3}{P_2} \right)^{(n-1)/n} - 1 \right]$$

$$\frac{1.25}{0.25} \times 14.7 \times 144 \times 1 \left[\left(\frac{100}{14.7} \right)^{0.25/1.25} - 1 \right]$$

$$5 \times 14.7 \times 144 \times (1.468 - 1) = 4950 \text{ ft-lb}$$

d. $\qquad \frac{4055}{5380} \times 100 = 75.3 \text{ percent}$

$$\frac{4950}{5380} \times 100 = 92 \text{ percent}$$

7.7 $\qquad \frac{1.24}{1.33} \times 100 = 93.2 \text{ percent}$

7.8

$$\text{hp} = \frac{w}{0.707}\left[c_p(t_2 - t_1) + \frac{V_2^2 - V_1^2}{50,000} + \frac{Z_2 - Z_1}{778}\right] + \left[\frac{w_j(t_0 - t_i) + R_e}{0.707}\right]$$

12,000 cfm = 200 cfs $P_1 = 14.0$ psia $T_1 = 60 + 460 = 520°R$

$$P_1 V_1 = wRT \qquad w = \frac{P_1 V_1}{RT_1} = \frac{14.0 \times 144 \times 200}{53.3 \times 520} = 14.55 \text{ lb.}$$

Velocity at entrance = 200/2.1 = 95.3 fps
Velocity at discharge = 200/0.4 = 500 fps

$$\text{hp} = \frac{14.55}{0.707}\left[0.24(624 - 520) + \frac{500^2 - 95.3^2}{50,000} + \frac{20}{778}\right]$$
$$+[677/60 \times (110 - 60)]/0.707$$
$$= 20.6(24.95 + 4.8 + 0.0256) + 797 = 1409 \text{ hp.}$$

Radiation loss $R_c = 0$.

7.9   Refer to Exhibit 7.9. One-fourth of the stroke is 4 in. Cylinder diameter is 12 in. The equation for work is given by

$$\text{Work} = P_1 V_1 \ln \frac{V_2}{V_1}$$

$$P_1 = 100 \times 144 = 14,400 \text{ psfa} \qquad V_1 = 0.785 \frac{12^2}{144} \times \frac{16}{12} \times \frac{1}{4} = 0.263 \text{ cu ft}$$

Work after cutoff = $100 \times 144 \times 0.263 \ln 4 = 5260$ ft-lb
Work up to cutoff = $100 \times 144(0.263 - 0) = \underline{3780 \text{ ft-lb}}$
Total = 9040 ft-lb

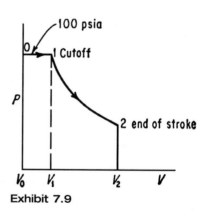

Exhibit 7.9

7.10   Assume efficiency of compressor as 80 percent, that of expander as 70 percent. Now refer to Exhibit 7.10. Work of compression

$$W_c = \frac{kP_1 V_1}{k-1}\left[1 - \left(\frac{P_2}{P_1}\right)^{(k-1)/k}\right]$$

$$W_c = \frac{1.4 \times 14.7 \times 144 \times 16,000}{(1.4 - 1) \times 33,000}\left[1 - 9.5^{0.40/1.4}\right] = 4100 \text{ hp}$$

Assume k is constant. Use 1.4 at 650°C. Gas is ideal at 75 psig.

$V_1 = 12{,}640 \times \left(\dfrac{15}{90}\right) \times (1660/520) = 6700$ cfm at 650°C and 75 psig. Work of expansion

$$W_e = \frac{1.4 \times 90 \times 144 \times 6700}{0.4 \times 33{,}000}\left[1 - \left(\frac{15}{90}\right)^{0.4/1.4}\right] = 2590 \text{ bhp}$$

Work to be supplied by motor is $-4100 + 2590 = -1510$ bhp.

Estimated motor size is 1750 hp.

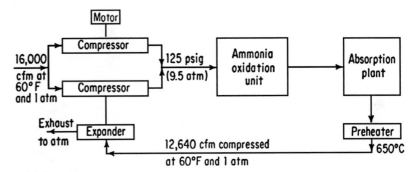

Exhibit 7.10

7.11 $$V_1 = V_d\left[1 + C - C\left(\frac{P_2}{P_1}\right)^{1/k}\right]$$

$$\frac{V_1}{V_d} = 1 + 0.08 - 0.08(94.7/14.7)^{1/1.4} = 0.777$$

$$\frac{V_1'}{V_d} = 1.08 - 0.08(87.35/7.35)^{1/1.4} = 0.61$$

Capacity decrease:

$1 - \{(0.61 \times 7.35)/(0.777 \times 14.7)\} = 0.617$ or 61.7 percent

Horsepower change:

$$W = \frac{kP_1V_1}{k-1}\left[1 - \left(\frac{P_2}{P_1}\right)^{(k-1)/k}\right]$$

$$W = \frac{1.4 \times 14.7 \times 0.777V_d}{0.4}[1 - (94.7/14.7)^{0.286}]$$

$$W = 40.0V_d(1 - 1.706) = -28.2V_d$$

$$W' = \frac{kP_1V_1'}{k-1}[1 - (87.35/7.35)^{0.286}]$$

$$W' = (1.4 \times 7.35)/0.4 \times 0.61V_d(1 - 2.024) = -16.1V_d$$

Decrease in theoretical horsepower required per unit volume is

$$1 - (16.1/28.2) = 0.43 \quad \text{or} \quad 43 \text{ percent.}$$

Increase in theoretical horsepower required per unit weight flow rate is

$$\frac{-16.1V_d}{-28.2V_d \times 0.383} = 1.49 \quad \text{or} \quad 49 \text{ percent.}$$

Temperature increase for 14.7 psia atmospheric pressure case is

$$T_2 = T_1 \left(\frac{94.7}{14.7}\right)^{(k-1)/k} = 540 \times 1.706 = 920°R \quad \text{or} \quad 460°F$$

and for 7.35 psia atmospheric pressure case is

$$T_2 = T_1 \left(\frac{87.35}{7.35}\right)^{(k-1)/k} = 540 \times 2.024 = 1.090°R \quad \text{or} \quad 630°F.$$

7.12 a.   Use k for air as 1.4; for methane (CH_4), 1.3.

For methane power $\qquad W_c = \dfrac{1.3P_1V_1}{0.3}[1 - 10^{2.3}] = -3.03P_1V_1$

For air $\qquad W_a = \dfrac{1.4P_1V_1}{0.4}\left[1 - \dfrac{259}{15}^{0.286}\right] = -4.4P_1V_1$

$(4.4 - 3.03)/3.03 = 0.45$ or 45 percent. At constant rpm and suction cfm, 45 percent more power is required for the air service.
Compressor not satisfactory.

b.   Steam 200 psia saturated has an enthalpy of 1199 Btu.

Steam at 800 psia saturated under constant entropy compression from 200 psia has an enthalpy of 1340 Btu. Theoretically $(1340 - 1199)/1340 = 0.041$ kwhr per lb of superheated steam at an enthalpy of 1340 Btu. Now let $X = $ lb superheated steam required per lb saturated steam. Then $1 - X = $ lb saturated condensate required per lb saturated steam. Then

$$1199 = (1 - X)689 + (1340X)$$

Solving for X, this is found to be 0.783, from which

$$0.041 \times 0.783 \times 100 = 3.21 \text{ kwhr per 100 lb}$$

800 psia saturated steam produced.

7.13 a.
$$h_a = \frac{kRT_1}{(k-1)M}\left[\left(\frac{P_2}{P_1}\right)^{(k-1)/k} - 1\right]$$

$$h_a = \frac{1.4 \times 1544 \times 580}{0.4 \times 29}[(47.2/14.7)^{0.286} - 1] = 43,400 \text{ ft}$$

b.
$$W_{ad} = \frac{1.4 \times 14.7 \times 144 \times 16,000}{0.4 \times 33,000}[1 - (47.2/14.7)^{0.286}] = -1435$$

$$E_{ad} = (1435/2100)100 = 68.5 \text{ percent}$$

c.
$$\frac{n-1}{n} = \ln\frac{\left\{1 - \left[\left(\frac{P_2}{P_1}\right)^{0.286} - 1\right]/0.685\right\}}{\ln(P_2/P_1)}$$

$$\ln 1.585/\ln 3.21 = 0.2/0.5065 = 0.395$$

From which n is found to be 1.65.

d.
$$E_p = \frac{n/n-1}{k/k-1} = \frac{2.54}{3.5}100 = 72 \text{ percent}$$

e.
$$h_f = \frac{E_p \times h_a}{E_a} = \frac{0.72 \times 43,400}{0.685} = 45,600 \text{ ft}$$

f.   For CO_2

$$h_f = \frac{nRT_1}{(n-1)M}\left[\left(\frac{P_2}{P_1}\right)^{(n-1)/n} - 1\right]$$

$$T_1 = 60 + 460 = 520°R \quad P_1 = 17.7 \text{ psia} \quad \text{Use } n = 1.65$$

$$\left(\frac{P_2}{P_1}\right)^{0.395} - 1 = \frac{45,600 \times 0.65 \times 44}{1.65 \times 1544 \times 520} = 0.981$$

$$\left(\frac{P_2}{P_1}\right)^{0.385} = 1 + 0.981 = 1.981$$

Thus, $P_2 = 17.7 \times 5.65 = 100 \text{ psia}$

g.
$$T_2 = T_1\left(\frac{P_2}{P_1}\right)^{(n-1)/n} = 520 \times 1.981 = 1030°R \quad \text{or} \quad 570°F$$

h.
$$W = \frac{1.65 \times 17.7 \times 144 \times 16,000}{0.65 \times 33,000 \times 0.72}[1 - 1.981] = -4280 \text{ bhp}$$

7.14   Specific speed $N_s = \dfrac{\text{rpm} \times \text{cfm}^{1/2}}{\text{SP}^{3/4}} = \dfrac{1200 \times 160,000^{1/2}}{16.0^{3/4}} = 60,000$

Diameter $d_s = \dfrac{d \times \text{SP}^{1/4}}{\text{cfm}^{1/2}} = \dfrac{40 \times 16^{1/4}}{160,000^{1/2}} = 0.20$ in.

7.15 $W_p = (20)(8.33)(1.5)(50) = 1.25 \times 10^4$ ft-lb/min

Actual work $= (1.25 \times 10^4/0.3) = 4.17 \times 10^4$ ft-lb/min

Volume of air required at ambient temperature assuming that the efficiency of the pump is based on reversible isothermal expansion is

$$W_{\min} \; NRT \ln (P_1/P_2) = P_1 V_1 \ln (P_1/P_2)$$

$1.25 \times 10^4 = (14.7)(144)(V_1) \ln (64.7/14.7)$, from which V_1 is found to be equal to 3.98 ft^3/min. Therefore, the actual volume is (3.98/0.3) or 13.27 ft^3/min.

Isentropic work for compressor:

$$-W = \frac{NRT_1^r}{r^{-1}} \left[\left(\frac{P_2}{P_1} \right)^{r-1/r} - 1 \right]$$

$$= \frac{(14.7)(144)(13.27)(1.4)}{0.4} \left[\left(\frac{64.7}{14.7} \right)^{0.4/1.4} - 1 \right]$$

$-W = 5.2 \times 10^4$ ft-lb/min

$\text{hp} = \dfrac{5.2 \times 10^4}{33,000} = 1.57$

Fuels and Combustion Products

PROBLEMS

8.1   Three-hundred pounds of carbon are burned with 65,000 cu ft of air initially at 14.7 psia and 300°F. Assuming that the carbon is pure, calculate the volumetric combustion analysis.

8.2   A coal contains 4 percent moisture, 23 percent volatile matter, 64 percent fixed carbon, and 9 percent ash and has a heating value of 14,100 Btu per lb. Determination of the carbon in the coal shows it to be 79 percent. The refuse removed from the ash pit of a grate-fired furnace using this coal contains 62 percent moisture (due to wetting down of the ashes by hose to lay dust), 3 percent volatile combustible matter, 11 percent fixed carbon, and 24 percent ash. Estimate the percent of the heating value of this coal lost in the furnace as unburnt combustible and the percent carbon fired that remains in the refuse.

8.3   Methane, CH_4, is burned in four times the theoretical air quantity, that is, the excess air is 300 percent. As a very crude approximation, it may be assumed that the specific heats of air, methane, and the combustion gases are equal. If the air is burned at 100°F, estimate the products of combustion temperature if the higher heating value of methane is 21,500 Btu per lb.

8.4   In the combustion of coal, what is the amount of heat liberated per pound of carbon in complete combustion to CO_2 and in combustion to CO?

8.5   A certain coal requires 8 lb of air per lb coal for perfect combustion. If in burning this coal at a rate of 1 ton per hr, 50 percent excess air is used, entering at 70°F and flue-gas temperature of 500°F, what is the heat loss due to this excess air?

SOLUTIONS

8.1 Refer to Chapter 8 in *License Review* and review the general considerations of combustion. Set up the theoretical combustion equation and note relative weights of each constituent. Refer to Table 8-1.

$$C + O_2 \rightarrow CO_2$$
$$12 + 32 \rightarrow 44$$

For theoretical combustion conditions and Table 8.1 the weight of air required to burn 1 lb of carbon is (2.67 + 8.78), or 11.45 lb. Theoretically, then, the total weight of air consumed by the 300 lb of carbon would be 300 × 11.45, or 3435 lb. Now let us see just what weight of air was injected into the firing chamber.

$$65,000 \times \frac{460+60}{460+300} \text{ (no pressure correction)} = 44,500 \text{ cu ft}$$

The weight of this air represents (44,500/379) 29 = 3400 lb. The actual weight of air fed to the combustion chamber is short by 3435 − 3400, or 35 lb. Under these conditions, incomplete combustion takes place and carbon monoxide appears in the flue gases. Now write out the theoretical combustion equation for CO formation.

$$2C + O_2 = 2CO$$
$$24 + 32 = 56$$

Total weight of carbon in the reaction is 300 lb. This weight must be distributed between that part reacted on to form CO_2 (let us call this part x) and that part reacted on to form CO (let us call this part $300 − x$). Now set up the following relationship in general terms first:

lb air per lb C to form $CO_2 \times x$ + lb air per lb C to form
CO$(300 − x)$ = wt of total air

$$\left(\frac{32}{12}\times1/0.2313\right)x+\left(\frac{32}{24}\times1/0.2313\right)(300-x)=3435$$
$$11.5x + 5.8(300 - x) = 3435$$
$$11.5x + 1740 - 5.8x = 3435$$
$$x = (3435 - 1740)/5.7 = 1695/5.7 = 298 \text{ lb}$$

Remembering that mole percent is equivalent to volume percent, calculate constituent breakdown on mole basis.

$$CO_2=298\times\frac{44}{12}\times\frac{1}{44}=24.8\text{ moles} \qquad 24.8/119 \equiv 20.8 \text{ percent}$$

$$CO = 2 \times\frac{56}{24}\times\frac{1}{28} = 0.2 \text{ mole} \qquad 0.2/119 \equiv 0.2 \text{ percent}$$

$$N_2 = 3435 \times 0.7687 \times\frac{1}{28} = 94 \qquad \text{moles } \frac{94}{119} \equiv 79 \text{ percent}$$
$$\text{Total moles} = 119 \text{ moles}$$

Note that no combustibles were present in the combustion chamber. All the carbon was reacted on. Some problems involve this complication and the next question will be so written.

8.2 Example 8.4 in *Mechanical Engineering: License Review* gives the ultimate analysis of the coal. The proximate analysis is presented here. Now proceed with solution.

The combustible matter in the refuse is not wholly coked fuel, as is evidenced by the presence of considerable volatile matter in it. Neither is it wholly uncoked coal, because the ratio of fixed carbon to volatile in the fuel, $64/23 = 2.78$, is not the same as that, $11/3 = 3.67$, obtained from the refuse. It is fair to assume that some wholly unburnt coal has dropped through the grate, the amount of this being measured by the volatile matter in the refuse. And note, too, that additional coal, coked completely in passing over the grate, has not had all of the carbon burnt out of it.

These are reasonable assumptions, because in the actual coking process there is little loss of volatile matter until a certain temperature is reached. However, when decomposition starts it is completed in a relatively narrow temperature range without much further heat supply and, on a furnace grate, in a relatively short time.

The volatile matter in the refuse is, obviously, a measure and proportional to the uncoked coal. Therefore, the loss in heating value due to uncoked coal is estimated as follows: basis 1 lb coal as fired.

$$0.09 \text{ lb ash} \times \frac{3}{24} \times \frac{100}{23} = 0.049 \text{ lb uncoked coal in refuse}$$

This represents a corresponding loss in heating value of $0.049 \times 14{,}000$, or 690 Btu.

The total fixed carbon in the refuse is made up of that due to uncoked coal as well as that from the coke present. The former, expressed per 100 lb of refuse, is $3 \times (64/23) = 8.35$, whereas—the total being 11—the difference, 2.65, is that corresponding to the coke present. Thus, the pounds of carbon in the refuse are

$$0.09(2.65/24) = 0.0099 \text{ per lb of coal}$$

and the corresponding heating value is

$$(0.0099/12)\,(97{,}000)\,1.8 = 145 \text{ Btu.}$$

Therefore, the total percent loss is

$$\frac{690+145}{14{,}100} \times 100 = 5.92 \text{ percent.}$$

The presence of moisture in the refuse presents no difficulties or complications, because ratios alone are used to transform from one basis to another.

If there is 0.049 lb of uncoked coal in the refuse, there is $0.049 \times 0.79 =$ 0.0387 lb of carbon in it. The total carbon unburnt is

$$0.0387 + 0.0099 = 0.0486 \text{ lb.}$$

This is equal to

$$(0.0486/0.79)(100) = 6.15 \text{ percent}$$

of the carbon in the coal.

8.3 Assume that specific heat for all gases in the products of combustion is 0.24 Btu/(lb) (°F), basis is 16 lb of methane, and that mixing is intimate. Heat absorbed is given by the equation $Q = wc_p\, \Delta t$. Now determine the weights of each of the flue gas components. Set up the theoretical equation for combustion of methane and balance it out.

$$CH_4 + 2O_2 = CO_2 + 2H_2O + 21,500 \text{ Btu per lb methane}$$

Products of combustion from the equation per 16 lb methane

$$CO_2 = 44 \text{ lb}$$
$$H_2O = 36 \text{ lb}$$
$$O_2 = 64 \times 3 = 192 \text{ lb}$$
$$N_2 = 64 \times (0.7687/0.2313) \times 4 = 850 \text{ lb}$$

Because each component will suffer the same temperature increase while absorbing its share of the total heat of combustion (of 16 lb methane)

$$44 \times 0.24 \times \Delta t + 36 \times 0.24 \times \Delta t + 192 \times 0.24 \times \Delta t + 850$$
$$\times 0.24 \times \Delta t = 21,500 \times 16$$
$$10.6\, \Delta t + 8.65\, \Delta t + 46\, \Delta t + 204\, \Delta t = 344,000 \text{ Btu}$$
$$\Delta t = 344,000/269.25 = 1280°F$$

Final temperature = 1280 + 100 = 1380°F.

8.4 To answer this question, we may take data on heating values directly from tables. The heating value of a fuel depends on the combustion conditions: temperatures of fuel and air at moment of ignition, and temperatures of products of combustion when the released Btu's are measured. Various values will be found in the literature, some based on 60°F, others on 32°F, etc. As a result, for instance, the heating value of CO ranges from 3900 to 4500 Btu.

The heat released by burning carbon to CO cannot be measured directly. It is taken as the difference between complete combustion of carbon to CO_2 and the combustion of CO to CO_2. The American Gas Association has published the following data based on temperatures of 32°F:

Carbon burned to CO_2 yields .... 169,686 Btu per mole C
Carbon monoxide burned to CO_2 yields .... −122,328 Btu per mole C
Difference is carbon burned to CO............. 47,358 Btu per mole C

Because the molecular weight of carbon is 12, it requires 169,686/12, or 14,140 Btu per lb of carbon in complete combustion to CO_2. For combustion to CO it liberates 47,358/12, or 3946 Btu per lb carbon.

8.5 Consider specific heat of air a variable expressed by

$$C_p = 0.239 + 4.138 \times 10^{-10}T^2$$

Weight of air involved is $2000 \times 8 \times (\frac{50}{100}) = 8000$ lb per hr excess air. This weight of air will absorb and carry away with it up the stack the amount of heat we are to determine. Also $T_1 = 460 + 70$, and $T_2 = 460 + 530$. Then

$$Q_{1-2} = W \int_{530}^{960} c_p dT = 8000 \int_{530}^{960} (0.239 + 4.138 \times 10^{-10})dT.$$

By integration the answer is found to be 8.25×10^5 Btu per hr.

Steam Power

OUTLINE

PROBLEMS

9.1 An acceptance test on a boiler shows that its induced-draft fan handles 600,000 lb per hr of flue gas at 290°F against a friction of 10 in. WG. The boiler is relocated to a spot where the barometric pressure is 24 in. Hg and not the original 30 in. Hg at sea level. With no changes made to the equipment except adjustments to the fan, and with gas weights and temperatures the same as before, what are the new volume and suction conditions for the fan design and relocation?

9.2 The allowable concentration in a certain boiler drum is 2000 ppm. Pure condensate is fed to the drum at the rate of 85,000 gallons per hour (gph). Make-up, containing 50 grains per gal of sludge-producing impurities, is also delivered to the drum at the rate of 1500 gph. Calculate the blowdown as a percentage of the boiler steaming capacity.

9.3 A process industry is studying the economics of a single-effect vs. a double-effect evaporation. Motive steam is available at 150 psia, dry and saturated; raw water from underground wells is in ample supply at a temperature that may be assumed constant at 60°F. Final evaporator product will be condensed at atmospheric pressure. Calculate the ratio of evaporator product to pounds of motive steam for (a) single-effect evaporation and (b) double-effect evaporation.

9.4 Superheated steam is generated at 1350 psia and 950°F. It is to be used in a certain process as saturated steam at 1000 psia. It is desuperheated in a continuous manner by injecting water at 500°F. How many pounds of saturated steam will be produced per pound of original steam?

9.5 By use of a suitable sketch, (a) Show the major equipment of, and indicate the principal circuits associated with, a boiling-water nuclear reactor power plant. (b) Name one advantage and one disadvantage of this cycle. (c) What is meant by negative temperature coefficient?

9.6 It is desired to estimate the fuel requirements for a nuclear reactor that is to operate at a thermal level of 500,000 kw continuously for a two-year period. The fuel is uranium containing $1\frac{1}{2}$ percent of the 235 isotope; fuel elements will be removed from the reactor for reprocessing when 1.0 percent of the 235 isotope has fissioned.

Preliminary calculations indicate that 800 megawatt-days of energy result from the fissioning of 1 g of U^{235}. On this basis, and using the enrichment and burnout percentages just given, determine the total weight in pounds of uranium (natural plus enriched) that must be put into the core.

9.7 It is desired to determine by calculation the enthalpy and the percent of moisture of the steam entering a condenser from a steam turbine. The turbine is delivering 20,000 kw and is supplied with steam at 850 psia and 900°F; the exhaust pressure is $1\frac{1}{2}$ in. Hg abs. The steam rate, when operating straight condensing, is 7.70 lb per delivered kwhr and the generator efficiency is 98 percent.

9.8 A 40,000-kw straight-flow condensing steam turbogenerator unit is to be supplied with steam at 800 psia and 800°F and is to exhaust at 3 in. Hg abs. The half-load and full-load throttle steam flows are estimated to be, respectively, 194,000 lb per hr and 356,000 lb per hr. The mechanical efficiency of the turbine is assumed to be 99 percent and the generator efficiency 98 percent.
  a. Compute the no-load throttle steam flow.
  b. Derive an equation for the heat rate of the unit expressed as a function of the output in kilowatts.
  c. Compute the internal steam rate of the turbine at 30 percent of full load.

9.9 The high-pressure cylinder of a turbogenerator unit receives 1,000,000 lb per hr of steam at initial conditions of 1800 psia and 1050°F. At exit from the cylinder the steam has a pressure of 500 psia and a temperature of 740°F. A portion of that steam is used in a closed feed-water heater to increase the temperature of 1,000,000 lb per hr of 2000 psia feed water from 350 to 430°F; the balance passes through a reheater in the steam-generating reheater and is admitted to the intermediate-pressure cylinder of the turbine at a pressure of 450 psia and a temperature of 1000°F. The intermediate cylinder operates nonextraction. Steam leaves this cylinder at 200 psia and 500°F. Calculate each of the following:
  a. Flow rate to the feed-water heater, assuming no subcooling
  b. Work done, in kilowatts, by the high-pressure cylinder
  c. Work done, in kilowatts, by the intermediate-pressure cylinder
  d. Heat added by the reheater.

9.10 In certain proposed nuclear power plants, the nuclear reactor together with specific accessory apparatus is enclosed in a containment vessel. In the event of rupture in the reactor coolant system, the containment vessel is expected to withstand the sudden increase in pressure. Describe your approach to the problem of specifying the maximum possible pressure for which the containment vessel should be designed. That is, how would you go about determining the maximum pressure that would develop following rupture of the reactor primary coolant system?

9.11 Water at 300°F is pumped into the boiler feed line, which is at a pressure of 1224 psia, and steam from the boiler leaves at 1200 psia and a total steam temperature of 700°F. Find the heat added to each pound of water passing through the boiler, assuming no heat losses whatsoever.

9.12 Determine the stack effect and i.d. fan needs for a fan located at the base of a stack. Static pressure of the fan may be consumed in overcoming resistance in the system ahead of the fan. Thus there is no pressure available at fan discharge. Ambient temperature may vary between 85°F and −30°F. Flow of stack gases will range between 175,000 cfm at 180°F and 250,000 cfm at 250°F. Calculation is to be based on summer ambient temperature of 85°F. In cold weather, efficiency of the stack improves because of the higher temperature difference. Fan discharges into the stack at 45°F and up.

9.13 For generation of steam at 90 psig, what is the advantage and percentage gain when boiler feedwater is heated from 50 to 210°F?

9.14 Choose a suitable feed-water regulator and combustion control for an industrial boiler serving the following loads: heating, 18,000 lb/h; process, 100,000 lb/h; miscellaneous uses, 12,000 lb/h. The boiler will have a maximum overload of 20 percent, and wide load fluctuations are expected at frequent intervals during operation. Pulverized-coal fuel will be used to fire the boiler.

9.15 A pump valve in our circulating system is provided with a spring having a spring rate of 65 lb/in. During the pumping cycle, the valve opens to its full limit of 1 in. The physical dimensions of the helical spring are 3.56 in. outside diameter, No. 1 W and M gauge (0.283 in.) wire diameter, free length of 4.25 in., and 6 total coils with ends squared and ground. When the valve is fully open, determine the total deflection, total load, and corrected maximum stress in the spring.

9.16 A steam turbine with a water rate of 80 takes steam at 150 lb/in.^{2} gauge, dry and saturated, and expands it to a backpressure of 25 lb/in.^{2} gauge at its outlet. (a) Determine the quality of the steam at the backpressure. (b) With a reciprocating pump, what would be the steam quality under the same conditions of throttle and backpressure? (c) Discuss energy transformations within each type of unit.

9.17 A processing plant operates a 5000-kW turbine having an engine efficiency of 73 percent. The initial steam conditions are 600 lb/in.^{2} abs, a total steam temperature of 600°F, and a backpressure of 1 in. Hg.
   a. Determine the turbine steam rate.
   b. Determine steam flow to the turbine throttle, lb/h.
   c. Determine the turbine heat rate if the turbine auxiliaries require 4000 lb/h of steam and their exhaust heats the feedwater to 160°F.
   d. Determine the station heat rate if the boiler efficiency is 80 percent and the electrically driven boiler auxiliaries require 100 kW.
   e. What is the overall station efficiency?

9.18   (a) What are the steps in sizing a control valve? (b) Size a steam control valve from the following situation: Inlet pressure of dry saturated steam at valve is 30 lb/in.^{2} abs. Flow rate is 1000 lb/h, with valve outlet pressure of 20 lb/in.^{2} abs. Valve type is straight-through (single-seat) throttling.

9.19   Industrial gases are often piped from water-sealed gas holders under suction to a processing plant some distance away. This gas is saturated with water vapor and is metered after pumping and the volume is automatically corrected to a base temperature and pipeline pressure as contracted for between gas supplier and customer. However, the water-vapor content is often neglected so that as delivery temperatures vary, actual volumes of gas and water vapor vary. If the customer has contracted for gas at 70°F delivered, its water-vapor content would be smaller than at a delivery temperature of 80°F. Because the customer will be billed for a volume of the saturated mixture at contract conditions, if the correction for water vapor is not made, he will pay for more gas than is actually delivered at the higher temperatures. On the other hand, the gas producer will be penalized for the lower-than-contract-temperature delivery.

Now, here's the situation. Barometric readings averaged at 29.21 in. Hg. Average gas pressure as determined by radial planimeter from the flow-meter recording chart was 15.04 lb/in.^{2} gauge. Average gas temperature was likewise determined as 51.57°F. Uncorrected volume of gas passing through the flowmeter for the period covered by the chart was 349,400 ft^3.

a.   Determine the corrected volume of gas passed through the flowmeter and corrected to 60°F and 30 in. Hg pressure and saturated with water vapor (i.e., contract conditions).

b.   What would be the difference if presence of water vapor were not corrected for?

c.   Who would be penalized, the gas producer or the customer?

9.20   A 12- by 16-in. single-acting steam engine takes steam for its full stroke at 100 psia and exhausts against a pressure (backpressure) of 16 psia. What is the net work done per revolution?

9.21   It is desired to build a compound engine with its low- and high-pressure cylinders double-acting. This engine is to develop 600 indicated hp when using steam having an absolute pressure of 150 psi and an absolute backpressure of 2 psi. The speed of the engine is to be 150 rpm and the piston speed in each cylinder is to be 750 fpm. If the cylinder ratio is to be 4 and the total ratio of expansion 12, find the size of the engine assuming a diagram factor of 0.80.

9.22   A steam engine develops 50 ihp with dry saturated steam supplied at 150 psia. The consumption is 1250 lb per hr. Calculate and determine the following: (a) Carnot efficiency, (b) Rankine efficiency, (c) actual thermal efficiency, and (d) engine efficiency.

9.23   Consider a heat engine that operates on the Carnot cycle with steam as the working substance. At the start of the adiabatic expansion, the pressure of the steam is 247.4 psia and the quality is 97 percent. The lowest temperature in the cycle is 100°F. Find (a) heat supplied, (b) heat rejected, and (c) cycle efficiency.

9.24 What would be the increase in power of a 12- by 18-in. steam engine running at 200 rpm noncondensing, if operated condensing with 26 in. Hg vacuum?

9.25 A power plant in a brewery is to generate 19,200 kw of electrical energy and be able to supply 40,000 lb per hr of steam for the unit operations within the plant. The process steam plus 30,000 lb per hr of steam for feed-water heating is extracted from the turbine at a point where the pressure is 25 psia. The turbine throttle conditions are 200 psia and 600°F total steam temperature. Condenser pressure is 2.0 psia. Compute the boiler capacity in pounds per hour of steam needed to serve these purposes at the throttle conditions.

9.26 A steam turbine carrying full load of 50,000 kw uses 569,000 lb of steam per hr. Engine efficiency is 75 percent and its exhaust steam is at 1 in. Hg abs and has an enthalpy of 950 Btu per lb. What is the temperature and pressure of the steam at the throttle?

9.27 By actual test after erection, how would you check the manufacturer's claimed operating conditions and state what you would expect the efficiency to be for a steam turbine?

9.28 A single-state turbine of the DeLaval type has a rotor 18 in. in diameter that revolves at 1200 rpm. The steam jet strikes the buckets at an angle of 20° with the plane of the rotor. It is desired to operate the rotor at 40 percent of the jet velocity. Its exhaust pressure is 14.7 psia and nozzle pressure is 100 psia. What steam temperature is necessary at admission to the nozzle?

9.29 Steam enters the fixed blade of a turbine with a relative velocity of 300 fps, a pressure of 70 psia, and 60° superheat. If the steam leaves the blade with a relative velocity of 800 fps and a pressure of 60 psia, find (a) its specific volume and entropy, (b) the blade loss in percent of available heat drop, and (c) the relative elocity at the exit if better blades were used so that the present loss is only 5 percent of available heat drop.

9.30 A 20,000-kw turbogenerator is supplied with steam at 300 psia and 650°F. Backpressure is 1 in. Hg abs. At best efficiency the combined steam rate is 10. Find (a) combined thermal efficiency, (b) combined engine efficiency, and (c) ideal steam rate.

9.31 If steam at a pressure of 200 psia and 500°F is supplied to a steam turbine, what percent increase in the efficiency of the Rankine engine cycle would result from lowering the backpressure from 15 psia (atmospheric) to 1 in. Hg?

9.32 Steam expands in a nozzle of a steam turbine from an initial pressure of 200 psia and a temperature of 450°F to a backpressure of 2 in. Hg abs. Weight of steam discharged is 7200 lb per hr. Friction loss in nozzle is 15 percent of theoretical (ideal) heat drop and radiation loss is 1 percent of ideal heat drop. Find quality of exhaust at nozzle exit.

9.33 An industrial plant operates a 5000-kw turbine having an engine efficiency of 73 percent. The initial steam conditions are 600 psia, 600°F, and the back-pressure is 1 in. Hg abs.
  a. Find the turbine steam rate.
  b. Find the pounds per hour to the turbine throttle.
  c. Find the turbine heat rate if the turbine auxiliaries require 4000 lb of steam per hr and their exhaust heats the feedwater to 160°F.
  d. Find the station heat rate if the boiler efficiency is 80 percent and the electrically driven boiler auxiliaries require 100 kw.
  e. What is the overall station efficiency?

9.34 An automatic extraction turbine operates with steam at 400 psia and 700°F at the throttle. Its extraction pressure is 200 psia and it exhausts at 110 psia. At full load 80,000 lb of steam per hr are supplied to the throttle and 20,000 lb per hr are extracted at the bleed point. What is the kilowatt output?

9.35 A 100-megawatt turbogenerator is supplied with steam at 1250 psia and 1000°F and the condenser pressure is 2 in. Hg abs. At rated load the steam supplied per hour is 1,000,000 lb, at zero load, it is 50,000 lb. What is the steam rate in pounds per kilowatt hour at $\frac{4}{4}$, $\frac{3}{4}$, $\frac{2}{4}$, and $\frac{1}{4}$ load?

SOLUTIONS

9.1 When a fan is required to handle air or gas at conditions other than standard (or any other basis), a correction must be made in the static pressure and horsepower. A fan is essentially a constant-volume machine and at a given speed in a given system the cfm will not materially change, regardless of density. The static pressure, however, changes directly with density. The static pressure must be carefully calculated for the specified conditions. Now for the problem at hand, assume a gas molecular weight of 28. Then the density correction factor is given as

$$\frac{24}{30} = 0.8$$

There is no temperature correction. Then the new suction condition is

$$10 \times 0.8 = 8 \text{ in. WG}$$

The new volume condition

$$\frac{600,000}{28} \times 379 \times \frac{460+290}{520} \times \frac{24}{24-8} = 17.6 \times 10^6 \text{ cu ft per hr}$$

9.2 We must first equate the grains per gallon of sludge to ppm to simplify the calculations further on.

$$\text{ppm} = \frac{50}{8.33 \text{ lb per gal} \times 7000 \text{ grains per lb}} = \frac{50}{8.33 \times 7000} = 855$$

To maintain a steady concentration in the boiler drum the sludge-forming impurities rate of feed through the normal arrangement must equate to the blowdown rate, or

$$\text{ppm to boiler} = \text{ppm in blowdown}$$

$$855 \times \text{gph make-up} = 2000 \text{ ppm} \times \text{gph blowdown}$$
$$855 \times 1500 = 2000 \text{ ppm} \times \text{gph blowdown}$$
$$\text{gph blowdown} = 855 \times 1500/2000 = 640$$

Boiler evaporation (steaming capacity)

$$= \text{feed} - \text{blowdown} = 85,000 + 1500 - 640 = 85,860 \text{ gph}$$

Finally, percent blowdown is

$$640/85,860 \times 100 = 0.747 \text{ percent}$$

9.3 See Exhibit 9.3. Use the steam tables. The heat losses are negligible.
a. For single-effect evaporation, heat in minus heat out is

$$1 \times (1194.4 - 330.5) = X(1150.4 - 28)$$

Solving for X, this is found to be equal to 0.77 lb of water evaporated per lb of steam supplied.
b. For double-effect evaporation, heat in minus heat out is

$$X(1150.4 - 180) = Y(1150.4 - 28)$$

$$Y = \frac{0.77(1150.4 - 180)}{(1150.4 - 28)}$$

$$= 0.665 \text{ lb water evaporated per lb steam}$$

Total evaporation per lb steam $= X + Y = 0.77 + 0.67$
$$= 1.44 \text{ lb per lb}$$

The problem as presented does not give the true picture for a double-effect evaporator because the second evaporator is not truly operating

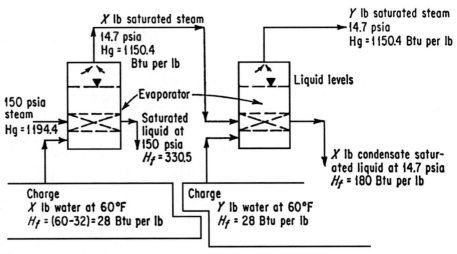

Exhibit 9.3

under a lower pressure than the first. The vapor coming from the first evaporator contains all the heat (no heat losses), latent in this case, put into it by the steam supplied to the first, and is reused as the steam supply to the second evaporator. In actual practice, the first effect could have been operated at a lower steam pressure, say 15 psia, because at lower pressures the latent heat is greater per pound. The second effect could have been operated at a vacuum of 18 in. Hg, heat being supplied by the vapor coming from the first effect at atmospheric pressure.

9.4   Enthalpy of steam at 1350 psia and 950°F = H_1 = 1465 Btu per lb
Enthalpy of saturated steam at 1000 psia = H_2 = 1191 Btu per lb
Enthalpy of water at 500°F (500 − 32) = H_3 = 488 Btu per lb

Let X = lb of 500°F water required, then

$$H_1 + XH_3 = (1+X)H_2 \qquad X = \frac{H_1 - H_2}{H_2 - H_3} = \frac{1465 - 1191}{1191 - 488} = 0.39.$$

Finally, 1 + 0.39 = 1.39 lb sat. steam produced per lb original steam.

9.5   a.   Exhibit 9.5a shows a simplified schematic of a nuclear power plant cycle. Fissioning (splitting) of uranium atoms in the reactor produces heat that is picked up by the cooling medium. The latter delivers this heat to the heat exchanger to generate steam to the turbine. The reactor and its cooling circuit must be heavily shielded to confine hazardous radiation within the reactor and cooling circuit.

b.   In nuclear power plants, the heat energy comes not from burning fuel but from splitting atoms. This fission process takes place in what used to be called a "pile," because that was a good description of the stack of graphite and uranium bricks under the west stands of Stagg Field at the University of Chicago where the first controlled chain reaction was produced. Now we call the device in which fission occurs a "reactor."

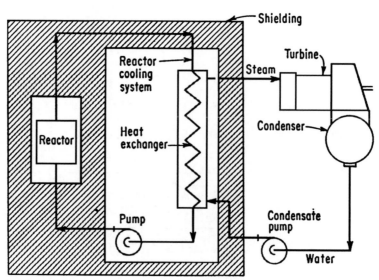

Exhibit 9.5a

A reactor can be likened to a furnace, not because it resembles one, but because it does the same job, releases heat. The nuclear reaction gives off so much heat that, even if we are not interested in making power, we must keep moving the heat to prevent the reactor from melting down. Heat in the coolant used can be transferred in a heat exchanger to generate steam that can spin the rotor of the turbogenerator. The heat exchanger and the reactor in the nuclear power plant are thus equivalent to the steam boiler and its furnace in the familiar steam power plant.

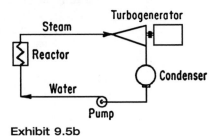

Exhibit 9.5b

Thus much of the nuclear power plant is familiar—turbine generator, condenser, pumps, and other auxiliary equipment. But there are some significant differences—all stemming from the unseen but seething atomic activity in the reactor.

The answer to part (a) is shown in Exhibit 9.5b. The coolant (which is also the moderator) picks up heat in the reactor core to form steam bubbles. The bubbles separate from the water at the water level and the steam leaves through the upper reactor tank nozzle. Feedwater enters the reactor tank below to pass up through the fuel elements in the core as coolant and moderator. Steam leaving the reactor will be moderately radioactive so that the steam equipment must be shielded.

Advantages	Disadvantages
1. Simple pumping needs. Much saving in pumping power over other systems.	1. Possible radioactivity of steam-power equipment and piping may require some shielding. Accessibility for maintenance and repair may be seriously hampered.
2. Note lower fuel temperatures. This is a definite advantage.	2. Radioactive particles in steam may require special shaft seals on turbines, pumps, valves, etc.
3. Simplified equipment layout. Less equipment needed. Note no heat exchanger needed as in Exhibit 9.5a.	3. Noncondensible gases removed from condenser may be radioactive.
4. Lower pressures required; thus lower pressure vessel expense.	

c. The great advantage of a boiling-water reactor is direct steam production without an external heat exchanger so that steam of given conditions can be produced with the same temperatures and pressures in the reactor. The reactor can be designed so that it has a *negative temperature coefficient* of reactivity. The formation of steam voids, as the power attempts to go up, will tend to shut down the reactor, or in other words, it has a large negative temperature coefficient.

9.6   For continuous operation, assume a plant factor of unity.

$$\frac{500 \text{ megawatts} \times 365 \text{ days} \times 2 \times 1}{800 \text{ megawatt-days}} = 457 \text{ g U}^{235} \text{ fissioned}$$

$$\text{Total weight} = \frac{457}{0.01 \times 0.015} \times \frac{1}{1000} \times 2.2046 = 6700 \text{ lb}$$

9.7   Refer to Exhibit 9.7 (Mollier diagram). The engine efficiency is

$$E_e = \frac{3413}{w_s(H_1 - H_c)} = \frac{3413}{7.7 \times (1453 - 898)} = 0.80 \text{ for ideal}$$

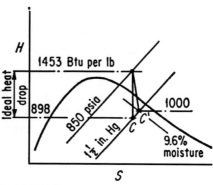

Exhibit 9.7

The Rankine engine efficiency for the steam turbine concerned is

$$\frac{0.80}{0.98} = 0.817 = \frac{H_1 - H_{c'}}{H_1 - H_c} \quad \text{and} \quad H_1 - H_{c'} = 0.817 \times 555$$

$$= 453 \text{ Btu per lb}$$

At the end of the actual expansion of the steam in the turbine $H_{c'}$ is

$$H_{c'} = 1453 - 453 = 1000 \text{ Btu per lb enthalpy}$$

where $H_{c'}$ crosses the pressure line of $1\frac{1}{2}$ in. Hg, moisture percent may be found to be 9.6 percent.

9.8   a.   See Exhibit 9.8a. Assume a straight-line rating characteristic; first determine the difference between full-load and half-load steam rates. This is $356,000 - 194,000 = 162,000$. Then $194,000 - 162,000 = 32,000$ lb per hr.

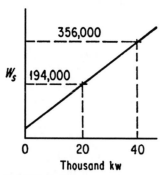

Exhibit 9.8a

b.   Refer to Exhibit 9.8b. Actual turbine efficiency is

$$E_t = \frac{3413}{w_s(H_1 - H_f)}.$$

$$\text{Heat rate} = \frac{3413}{E_t} \text{ Btu per kwhr} = w_k(H_1 - H_f)$$

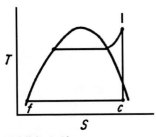

Exhibit 9.8b

Actual steam rate in pounds per kilowatt hour is $w_k = w_s/kw$.

$$w_s = 32,000 + \frac{162,000}{20,000} \times kw = 32,000 + 8.1 \; kw$$

$$w_k = \frac{32,000}{kw} + 8.1$$

Now with $H_1 = 1398$ and $H_f = 83$, then $H_1 - H_f = 1315$.

Finally,

$$\text{Heat rate} = \frac{1315 \times 32,000}{kw} + [1315(8.1)]$$

$$= 10,650 + \frac{42,100,000}{kw}$$

Let us see how the steam rate and heat rate compare for the quarter-load points.

At full load: $w_k = 8.9$ lb per kwhr   Heat rate = 11,700 Btu per kwhr

At $\frac{3}{4}$ load: $wk = 9.17$ lb per kwhr   Heat rate = 12,080 Btu per kwhr

At $\frac{1}{2}$ load: $w_k = 9.7$ lb per kwhr   Heat rate = 12,770 Btu per kwhr

At $\frac{1}{4}$ load: $w_k = 11.3$ lb per kwhr   Heat rate = 14,870 Btu per kwhr

c. For engine and turbine combined

$$E_e = \frac{3413}{w_k(H_1 - H_c)}.$$

Also

$$H_1 = 1398 \quad H_c = 912 \quad H_1 - H_c = 486$$

We saw that

$$w_s \text{ at full load} = 356,000$$
$$w_s \text{ at no load} = 32,000.$$

For the full-load range the total change is

$$356,000 - 32,000 = 324,000.$$

$$wk \text{ at 30 percent load } \frac{32,000 + 0.30(324,000)}{0.30 \times 40,000} = 10.87 \text{ lb per kwhr}$$

$$E_i = \frac{3413}{10.78 \times 486} = 0.651 \text{ for engine and turbine combined}$$

If we call the internal efficiency of the turbine (not including the friction loss) E_i, then

$$E_i = \frac{2545}{w_a(H_1 - H_c)}.$$

Thus

$$E_i = \frac{E_e}{\text{mechanical efficiency} \times \text{generator efficiency}}$$

$$E_i = \frac{0.651}{0.99 \times 0.98} = 0.672$$

The actual steam rate in pounds per horsepower is

$$w_a = \frac{2545}{E_i(H_1 - H_c)} = \frac{2545}{0.672 \times 486} = 7.8.$$

9.9   See Exhibit 9.9. Use the steam tables and Mollier diagram to find the various enthalpies. Interpolate as required. Neglect pressure drop through feedwater heater in lines and other heat-exchange apparatus except as noted.

a.   Steam flow rate to feed-water heater: This is merely a heat balance.

Flow to heater $(H_2 - H_7)$ = water rate $(H_6 - H_5)$

$$\text{Flow to heater} = \frac{1 \times 10^6 (409 - 324.4)}{(1379.3 - 449.4)} = 91,100 \text{ lb per hr}$$

b.   Work done by high-pressure cylinder:

$$\text{Work done} = \frac{1 \times 10^6 (1511.3 - 1379.3)}{3413} = 38,700 \text{ kw}$$

c.   Work done by intermediate-pressure cylinder:

$$\text{Work done} = \frac{(1 \times 10^6 - 91.1 \times 10^3)(1521 - 1269)}{3413} = 67,150 \text{ kw}$$

d.   Heat added by reheater:

$$\text{Heat added} = 908,900(1521 - 1379.3) = 128.9 \times 10^6 \text{ Btu per hr}$$

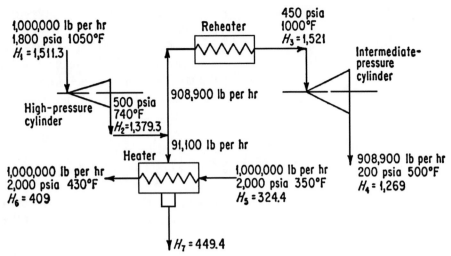

Exhibit 9.9

9.10 Assume the type of enclosure to be a "reactor only," one not housing the turbogenerator. By housing the turbine in a more nearly conventional building, it renders the plant more amenable to future modifications—an important consideration in a development project such as nuclear power plants are today.

From the nuclear safety standpoint, too, the "reactor only" style of enclosure offers other advantages and simplifies the problem of pressure calculation. The danger of a turbine explosion is minimized with the turbine outside, rotating in a plane not intersecting the containment vessel. Then, too, the danger of damage to the enclosure by explosion of the hydrogen used to cool the generator is eliminated. With less lubricating oil within the enclosure the fire safety is improved.

The enclosure design is usually based on adequacy for pressures that could be developed in severe terminal accidents. However, experience with present installations indicates the possibility of occurrence of any such accident would be extremely remote.

The internal pressure within the containment vessel (enclosure) in the event of a major accident could be created by contributions from three different potential sources of energy:

a. The pressurized hot water in the reactor and in those of its auxiliaries that are not separated from it by a solid barrier. In the Dresden Nuclear Power Station outside Chicago, Illinois, a reactor rupture would expose 188 tons of pressurized hot water at 1000 psig in a boiling condition. Volume of containment vessel is that from a 190-ft-diameter sphere.

b. A nuclear excursion.

c. Chemical reaction between reactor components, that is, between water and the zirconium cladding in which the uranium oxide fuel is encased.

The hot-water contribution is the most important of these. Any nuclear and chemical contributions that would be significant in comparison are improbable.

Accordingly, the design pressure should be determined by the pressure that would be created by release and partial flashing of water to steam of the pressurized hot water contained in the reactor and auxiliaries with no significant nuclear or chemical energy contributions. For the Dresden installation at its rated operating pressure (1000 psig) and rated thermal level (630 megawatts) the pressure expected to develop is about 25 psig.

Design pressure normally is 30 psig with a 1.25 times design pressure for pneumatic testing.

The basis for design temperature not required in this problem should not go without some word. The equilibrium temperature corresponding to a pressure of 25 psig design pressure plus 100°F brings the design temperature up to slightly over 300°F.

9.11   Initially the water is a subcooled liquid. The heat added to each pound is the difference in the initial and final state points. The heat equivalent of work done on each pound of liquid by the feedwater pump must be included. Enthalpy of the liquid initially is

$$\text{Enthalpy at } 300°F + [A(P_1 - P_a)]\,\bar{V}_a$$

$$269.5 + \left[\left(\frac{1}{778}\right)(144)(1224 - 67)\right](0.0174) = 273.2 \text{ Btu per lb.}$$

Subscript a denotes the condition on the saturated liquid line at 67 psia (the saturation pressure for 300°F). The final enthalpy at the superheater outlet is 1310.6 Btu per lb from steam tables or Mollier diagram. Thus, heat added per lb = 1310.6 − 273.2 = 1037.4.

9.12   Theoretical draft or static at the base of stack calculated from the difference in weight between the column of gas and similar column of ambient air is

$$\text{Static pressure} = (0.256) \times b \times H \times \left(\frac{1}{T_{amb}} - \frac{1}{T_m}\right)$$

$$= (0.256) \times 28.75 \times 80 \times \left(\frac{1}{545} - \frac{1}{640}\right)$$

$$= 0.108 \text{ in. water column}$$

where

b = barometric pressure at 1100 ft elevation, in. Hg
H = stack height, 80 ft; stack diameter = 11 ft
T_{amb} = 460 + 85 = 545°R
T_m = 460 + 180 = 640°R

The gas flow through the stack depends on the losses that must be overcome, including friction loss through the stack itself.

Airflow	175,000 cfm	250,000 cfm
Stack velocity	1850 fpm	2640 fpm
Velocity pressure @ 130°F avg	0.201 in. wc	0.422 in. wc
Elbow velocity	3300 fpm	4800 fpm
Velocity pressure @ 130°F avg	0.6 in. wc	1.25 in. wc
Stack friction loss	0.045 in. wc	0.075 in. wc
Stack exit loss	Negligible	Negligible
Elbow loss	0.164 in. wc	0.34 in. wc
Stack entrance loss	Negligible	Negligible
Total losses	0.209 in. wc	0.415 in. wc
Plus 10 percent safety factor	0.021 in. wc	0.042 in. wc
Required stack draft	0.230 in. wc	0.457 in. wc
Available stack draft	0.108 in. wc	0.108 in. wc
Minimum in. wc to be supplied by fan at stack	0.122 in. wc	0.349 in. wc

Available stack draft alone cannot remove gases from the furnace. Fan must be speeded up and larger fan motor provided to give the additional static needed.

9.13 There is a definite advantage. A pound of dry saturated steam at 90 psig contains 1187 Btu above 32°F, and when the feedwater is at the temperature of 50°F, or (50 − 32), 18°F above 32°F, for conversion into steam at the stated pressure each pound must receive 1187 − 18 or 1169 Btu. When feedwater is at 210°F, each pound of feedwater evaporated requires 210 − 50 or 160 Btu less, or only (1169 − 160)/(1169) × 100 or 86.3 percent as much heat conversion into steam, and the gain would be 13.7 percent.

9.14 Determine the required boiler rating by first finding the sum of the individual loads on the boiler, or 18,000 + 100,000 + 12,000 = 130,000 lb/h. With a 20 percent overload, the boiler rating must be 1.2 × 130,000 = 156,000 lb/h. With an additional reserve capacity to provide for unusual loads, the rated boiler capacity should be 1.1 × 156,000 = 171,500 lb/h, say 175,000 lb/h for selection purposes.

Choose the type of feedwater regulator to use. A boiler in the 75,000 to 200,000 lb/h range can use a relay-operated regulator with one or two elements when the load fluctuations are reasonable. With wide load swings the relay-operated three-element regulator is a better choice. In addition, because the boiler will encounter wide load swings, a three-element regulator is a wise choice and a safe one, too.

Choose the type of combustion-control system. A stream-flow-airflow type of combustion-control system would probably be the best for the fuel and load conditions in this plant. This is the type of combustion-control system to use.

Any control system selected for a boiler should be checked out by studying the engineering data available from the control system manufacturer.

Excess pressure ahead of the feedwater regulator should be at least 50 lb/in.² and should be controlled by regulation of the feed pump. Use excess pressure valves only when excess pressure varies more than plus or minus 30 percent. Where drum level is unsteady owing to high solids concentration or boiler feed or other causes, use next higher-class feed regulator. See Kallen, *Handbook of Instrumentation and Controls*, McGraw-Hill.

9.15   Initial compression = 100/65 = 1.538 in

$$\text{Total deflection} = 1.538 + 1.0 = 2.538 \text{ in.}$$
$$\text{Valve opening load} = 65 \times 1.0 = 65 \text{ lb}$$
$$\text{Total spring load} = P = 100 + 65 = 165 \text{ lb}$$
$$\text{Mean diameter of spring} = 3.56 - 0.283 = 3.277 \text{ in.}$$
$$\text{Spring index} = D/d = 3.277/0.283 = 11.59 \text{ and Wahl}$$
$$\text{correction } K$$

where

$$K = (4C - 1)/(4C - 4) + (0.615)/(C) = 1.124$$
$$S_{vc} = (8PDX)/(\pi d^3)$$
$$S_{vc} = (8 \times 165 \times 3.277 \times 1.124)/(3.1416 \times 0.283^3) = 68{,}000 \text{ lb/in.}^2$$

9.16   Refer to Exhibit 9.16. This is an H-S diagram.

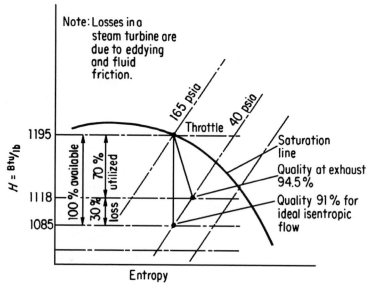

Exhibit 9.16

a.   Steam turbines with water rates of 60 to 80 transform 70 to 75 percent of the heat for energy. Turbines with higher rates transform 80 to 85 percent of the heat for energy. Reciprocating pumps or engines transform 50 percent of the heat available. The above data are based on practical tests.

   Now, from Exhibit 9.16, steam tables, and Mollier diagram and assuming 70 percent utilization of heat,

$$(1195 - 1085)(0.70) = 77$$
$$1195 - 77 = 1112 \text{ Btu/lb enthalpy at exhaust}$$

From Mollier diagram as shown, the quality of exhaust steam is found to be 91 percent. Ideal heat drop is equivalent to 100 percent availability of energy.

b.   For the reciprocating unit the ideal heat drop is still $(1195 - 1085) = 110$ Btu/lb steam. Percentage utilized is 50, so that the actual heat drop is $(1195 - 1085)(0.50) = 55$ Btu/lb steam. Then as before the quality is found to be 94 percent.

c. The higher the thermal efficiency the lower the steam quality after expansion through the equipment. Also, the higher the water rate the lower the steam quality at the exhaust. Exhaust from a reciprocating driver is high in steam quality because of wire drawing through the valve ports. Losses in steam turbines are due to eddying and fluid friction.

9.17 a. Draw the *TS* diagram for the process. See Exhibit 9.17.

Let $P_1 = 600$ lb/in.^{2} abs $p_2 = 1$ in. Hg $\eta_e =$ engine efficiency

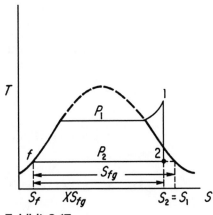

Exhibit 9.17

$S =$ entropy $t_1 = 600°F$ $\eta_b =$ boiler efficiency
$h =$ Btu/lb steam $X =$ quality of steam, percent

From steam tables at $P_1 = 600$ lb/in.^{2} abs and $t_1 = 600°F$

$h_1 = 1290.9$ $S_1 = 1.5334$ $S_1 = S_2$ isentropic expansion

At 1 in. hg(approximately 0.5 lb/in.^{2} abs)

$$h_f = 47.5 \quad S_f = 0.0924 \quad t_f = 79°F$$
$$h_{fg} = 1048 \quad S_{fg} = 1.9434$$

Now $S_2 = S_f + X\,S_{fg}$ or $1.5334 = 0.0924 + X(19{,}434)$, from which $X = 0.74$, or steam quality at point 2 (Exhibit 9.17) is 74 percent. Then, $h_2 = h_f + X\,h_{fg}$ or $h_2 = 47.5 + 0.74 \times 1048 = 823.5$ Btu/lb steam.
    Note: h_2 and percent moisture can be obtained directly from the Mollier diagram. The above arithmetic method is more accurate and is given in case the Mollier diagram is not available.

Then ideal work $W_i =$ ideal heat drop $= h_1 - h_2 = 1290.9 - 823.5 = 467.4$ Btu/lb.

$$\text{Rankine steam rate} = \frac{3413}{h_1 - h_2} = \frac{3413}{476.4} = 7.3 \text{ lb/kWh}$$

Actual work $W_a = \eta_e X W_i = 0.73 \times 467.4 = 341$ Btu/lb steam
Turbine steam rate $= 3413/W_a = 3413/341 = 10$ lb/kWh

b. Steam rate to throttle $= 10 \times 5000 = 50{,}000$ lb/h

c.   Input = (throttle steam + auxiliary steam)$(h_1 - h_f)$. Then from steam tables h_f at 160°F = 127.9.

$$\text{Heat rate} = \frac{(50,000 + 4000)(1209.9 - 127.9)}{5000 \text{ kWh}}$$
$$= 12,560 \text{ Btu/kWh}$$

d.   Turbine heat rate per net kWh, where net kWh = 5000 − 100

$$\text{Turbine heat rate} = \frac{62,800,000}{4900} = 12,816 \text{ Btu/kWh}$$

The turbine heat rate per net kWh represents boiler output.

Then the station heat rate = (turbine heat rate/boiler efficiency)
= 12,816/0.8 = 16,020 Btu/kWh

e.   Station efficiency = (3413/station heat rate) = 3413/16,020
= 0.213 or 21.3 percent

9.18   **a.**   Select type of valve in accordance with practice.

Compute the critical pressure P_c.

Determine fluid density p using steam tables.

Calculate valve coefficient C_v.

Calculate $C_v/0.8$.

Select valve size, using manufacturer's rating tables.

b. $P_c = P_1 \times 0.58 = 17.4 \text{ lb/in.}^2$ abs.
$P_2 = 20 \text{ lb/in.}^2$ abs.

Thus, P_2 is greater than P_c. Therefore flow is noncritical. ρ is based on P_2 conditions. Steam tables give the specific volume of the steam at $P_2 = 20 \text{ lb/in.}^2$ abs as 20.11 ft³/lb. Thus, $\rho = (1)/(20.11) = 0.05 \text{ lb/ft}^3$. Calculate C_v from the following standard formula:

$$C_v = \frac{W}{(63.5)(P_1 - P_2)\rho}$$

where

C_v = valve flow coefficient
P_1 = valve inlet pressure, lb/in.² abs
P_2 = valve outlet pressure, lb/in.² abs
ρ = vapor (or gas) density, lb/ft³
W = vapor or gas flow rate, lb/h

$$C_v = \frac{1000}{(63.5)(30 - 20)(0.05)} = 31.5$$

From manufacturers' ratings select a 2-in. valve straight-through throttling single seat. For normal operation, maximum operating flow should not be greater than 80 percent of the maximum possible flow (C_v for the valve in question). Flow values of less than 10 percent should not be used; that is, calculated flow rate divided by C_v should not be less than 10 percent.

9.19 a. Average absolute gas pressure

$$15.04 + (29.21 \times 0.4912) = 29.39 \text{ lb/in.}^2 \text{ abs}$$

Gas volume reduced to standard conditions

$$V_c = 349,400 \times \text{correction factor}$$

$$V_c = 349,400 \times \frac{29.39 - 0.1888}{14.69 - 0.256} \times \frac{519.6}{51.57 + 459.6} \qquad (9.1)$$

$$V_c = 349,400 \times 2.056 = 718,366 \text{ ft}^3$$

This is the volume of gas reduced to standard gas saturated with water vapor at 60°F and 30 in. Hg and corrected for water vapor.

b. If the presence of water vapor were not corrected for

$$V = 349,400 \times \frac{29.39}{14.69} \times \frac{519.6}{511.17} = 709,981 \text{ ft}^3$$

or a difference of $718,366 - 709,981 = 8385 \text{ ft}^3$.

c. The gas producer would be penalized.
Explanation of equation 9.1.

$$V_c = V \frac{p - w}{B - W} \times \frac{T}{t}$$

where

V = observed flowmeter reading, ft^3
V_c = volume of standard gas saturated at 30 in. Hg and 60°F
p = observed absolute gas pressure, lb/in.^{2} abs
w = vapor pressure of water at t, lb/in.^{2} abs
B = standard barometer equivalent, 14.69 lb/in.^{2} abs
W = vapor pressure of water at 60°F, lb/in.^{2} abs
T = °Rankine
t = observed absolute gas temperature, °R

9.20 See Exhibit 9.20. V_1 is equal to zero. At the end of the full stroke V_2 is equal to

$$V_2 = 0.785 \left(\frac{12}{12}\right)^2 \times \left(\frac{16}{12}\right) = 1.05 \text{ cu ft}$$

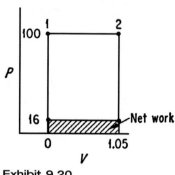

Exhibit 9.20

Work on the forward stroke is between points 1 and 2. Thus,

$$W_{1-2} = 100 \times 144(1.05 - 0) = 15,125 \text{ ft-lb.}$$

On the return stroke work is done on the steam; this is W_{2-1}.

$$W_{2-1} = 16 \times 144(0 - 1.05) = -2420 \text{ ft-lb work done on steam}$$

New work per revolution is $15,125 - 2420 = 12,705$ ft-lb.

9.21   The length of stroke is 750 fpm/$(2 \times 150) = 2.5$ ft, or 30 in. The ideal mean effective pressure is

$$P_m = \frac{\text{area of ideal } PV \text{ diagram}}{\text{length of diagram}}.$$

This is also

$$P_m = P_1 \left(\frac{1 + \ln r_e}{r_e} \right) - P_3.$$

Also the ratio of expansion r_e is equal to 1/cutoff. In the problem at hand

$$P_m = 150 \left(\frac{1 + \ln 12}{12} \right) - 2 = 150 \times 0.290 - 2 = 41.5 \text{ psia.}$$

With the diagram factor being 0.8, the actual mean effective pressure (mep) is

$$\text{Mep} = 41.5 \times 0.80 = 33.2 \text{ psia.}$$

Now use the basic equation for hp = $PLAN/33,000$ and arrange to solve for A.

$$A = \frac{600 \times 33,000}{2 \times 33.2 \times 2.5 \times 150} = 795 \text{ sq in.}$$

This value of 795 sq in. gives a valve of 32 in. for the diameter of the low-pressure cylinder. The diameter of the high-pressure cylinder is solved by $32/\sqrt{4} = 16$ in. The machine would be a 16 by 32 by 30 in. Note diagram factor is ratio of area of actual indicator card to area of ideal indicator card.

9.22   Let the following nomenclature apply:

T_1 = steam temperature at inlet pressure, °R
T_2 = steam temperature at exhaust pressure, °R
H_1 = enthalpy of steam at inlet pressure, Btu per lb
H_2 = enthalpy after isentropic expansion to exhaust (back) pressure, Btu per lb
H_3 = enthalpy of saturated liquid at exhaust pressure, Btu per lb
w_a = actual water rate of engine, lb per ihp-hr

Obtain all values but w_a from steam tables.

a.   Carnot efficiency $= \dfrac{T_1 - T_2}{T_1} = \dfrac{(460+358)-(460+219)}{460+358}$, or 17 percent

b.   Rankine efficiency $= \dfrac{H_1 - H_2}{H_1 - H_3} = \dfrac{1194 - 1034}{1194 - 188}$, or 15.9 percent

c.   Actual thermal efficiency $= \dfrac{2545}{(H_1 - H_3)(1/w_a)}$

$\dfrac{2545}{(1194-188)\times 1/(1250/50)}$, or 10.1 percent

d.   Engine efficiency $= \dfrac{\text{actual thermal efficiency}}{\text{Rankine efficiency}} = \dfrac{0.101}{0.159}$
$= 0.635$, or 63.5 percent

9.23   a.   Heat supplied $= T_1 (\Delta S)_t$. But first we must determine ΔS.

$S_e = S_f + (x_e S_{fg}) = 0.5668 + (0.97 \times 0.9602) = 1.498$
$S_d = S_a = S_f = 0.5668$

Heat supplied $= T_1(\Delta S)_t = (860)(1.498 - 0.567) = 801$ Btu per lb

b.   Heat rejected $= T_2(\Delta S)_t = (560)(1.498 - 0.567)$
$= 521$ Btu per lb

c.   Efficiency $= \dfrac{T_1 - T_2}{T_1} = \dfrac{400 - 100}{860} = 0.349$, or 34.9 percent

9.24   A 26-in. Hg vacuum is a pressure of

$$(26 \times 0.491) = 12.76 \text{ psi}$$

less than atmospheric pressure. If the backpressure when operating non-condensing is 2 psi above atmospheric pressure, the reductions of backpressure from operating condensing would be

$$(2 + 12.76) = 14.76 \text{ psia}$$

The power would be increased

$$\frac{14.76(12\times12\times0.785)(18)(2)(200)}{12\times 33,000} = 30.35 \text{ ihp.}$$

Note that engine is double-acting and the sectional area of piston rod on crank end was not deducted because complete information was not given in the problem.

9.25 Refer to Exhibit 9.25.

$$p_1 = \text{throttle pressure} = 200 \text{ psia}$$
$$t_1 = \text{throttle temperature } 600°F$$
$$p_2 = \text{extraction pressure} = 25 \text{ psia}$$
$$p_3 = 25 \text{ psia}$$
$$p_4 = \text{condenser pressure} = 2.0 \text{ psia}$$

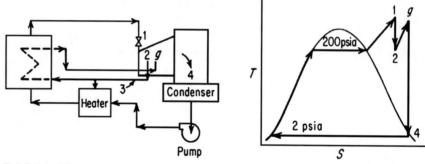

Exhibit 9.25

From Mollier diagram,

$$h_1 = 1322.1 \text{ Btu per lb}$$
$$h_2 = 1134.4 \text{ Btu per lb}$$
$$h_3 = 1134.4 \text{ Btu per lb}$$
$$h_4 = 974 \text{ Btu per lb}$$

Turbine power needed to generate, assuming generator efficiency = 95 percent:

$$19,200 \times 3413 \times 1/0.95 = 6.90 \times 10^7 \text{ Btuh}$$

If turbine efficiency is assumed to be 70 percent, then the isentropic power required is

$$6.90 \times 10^7/0.7 = 9.86 \times 10^7 \text{ Btuh.}$$
$$\text{Steam to process} = 40,000 \text{ lb per hr}$$
$$\text{Steam from feedwater heating} = 30,000 \text{ lb hr}$$
$$\text{Total} = 70,000 \text{ lb pr hr}$$

Set up an energy balance in this way:

$$(70,000 + \text{lb per hr through turbine})(1322.1 - 1134.1)$$
$$+ (\text{lb per hr through turbine})(1134.4 - 974) = 9.86 \times 10^7 \text{ Btuh}$$

from which lb per hr through turbine = 216,982 lb per hr.

$$\text{Steam to process} = 40,000 \text{ lb per hr}$$
$$\text{Steam to feedwater heating} = 30,000 \text{ lb per hr}$$
$$\text{Steam through turbine} = 216,892 \text{ lb per hr}$$
$$\text{Boiler capacity needed} = \text{sum of above} = 286,892 \text{ lb per hr}$$

9.26 See Exhibit 9.26. Engine efficiency is given in accordance with

$$E_e = \frac{3413}{w_s(H_1 - H_2)}$$

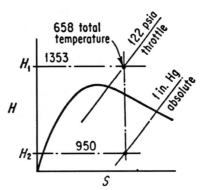

Exhibit 9.26

First we see that w_s may be found by a simple ratio from given data.

$$w_s = 569,000/50,000 = 11.38 \text{ lb per kwhr}$$
$$H_1 - H_2 = 3413/(11.38 \times 0.75) = 403 \text{ Btu}$$
$$H_1 = 403 + 950 = 1353 \text{ Btu per lb}$$

Referring to the skeleton Mollier diagram (Exhibit 9.26) and following the isentropic path upward, we see that the vertical path intersects the enthalpy line of 1353 Btu per lb at 122 psia and 658°F total steam temperature.

9.27 An acceptance test should be made of a steam turbine to determine whether the guaranteed value of the steam consumption has been met. The test should be made under the operating conditions that were specified for the machine guarantee conditions. The steam supply (assumed 100,000 lb per hr) should be noted from a steam flow meter. The system steam pressure (assumed 450 psig) is noted by pressure gauge (calibrated). The steam temperature is noted from a calibrated thermometer or indicator. The exhaust or backpressure (29 in. Hg) is noted by vacuum gauge (calibrated). The load (12,000 kw-hr) is noted from voltmeter and ammeter. The actual steam consumption is about 8.35 lb per kwhr, while the ideal steam consumption would be about 6.82 lb per kwhr, which would give an efficiency of about 82 percent. Note it would not be remiss to repeat that all flow-measuring and temperature- and pressure-indicating instruments must be carefully calibrated before a test is conducted.

9.28 See Exhibit 9.28.

Let v_b be rotor velocity in feet per second. Then

$$v_b = (2\pi)\left(\frac{18}{2}\right)\left(\frac{1}{12}\right)(12,000/60) = 942 \text{ fps}$$
$$v_1 = (1/0.4)(942) = 2355 \text{ fps}.$$

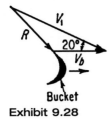

Exhibit 9.28

Drop through blading is isentropic across terminal conditions. Proceeding,

$$\frac{v_1^2}{2gJ} = \text{ideal heat drop} \times 0.80 = \text{actual heat drop.}$$

Ideal heat drop $= (2355)^2/(64.4 \times 778 \times 0.80) = 138.5$ Btu per lb. Following the path of the expansion on the Mollier diagram between 100 psia and 14.7 psia, we obtain close to 138.5 Btu per lb. This gives us a temperature of 327.8°F.

9.29 Using a form of the general energy equation and referring to the skeleton Mollier diagram (Exhibit 9.29),

a. $$v_1^2 - v_2^2 = 50,000(H_1 - H_2)(1 - y)$$

where $H_1 - H_2$ is the ideal heat drop and y is an expression of turbine losses as a decimal less than one.

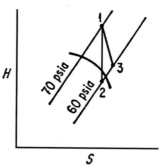

Exhibit 9.29

Then for the actual heat drop

$$v_1^2 - v_2^2 = 50,000(H_1 - H_3)$$
$$300^2 - 800^2 = 50,000(1212 - H_3)$$

From this H_3 is found to be equal to 1201.5 Btu per lb. Referring to Exhibit 9.29 and knowing H_3 and the pressure condition of 60 psia, we locate this point on the Mollier diagram. The final entropy is 1.6750 and the specific volume may be determined by the quality equation $v = xv_f + (1 - x)v_f$. Thus, the specific volume is 7.689 cu ft per lb.

b. Blade loss is $(H_3 - H_2)/(H_1 - H_2)$.

$(1201.5 - 1199.3)/(1212.5 - 1199.3) = 0.1675$, or 16.75 percent

c. For better blades we substitute for the y the value 0.05, insert in the first equation given containing the term y, and solve for v_2. This is found to be 867 fps. Note $S_1 = S_2 = 1.6721$.

9.30 a. Combined thermal efficiency (see Exhibit 9.30):

$$\frac{3413}{w_e(H_1 - H_3)} = \frac{3413}{10(1340.6 - 47.06)} = 0.264, \text{ or } 26.4 \text{ percent}$$

b. Combined engine efficiency: ideal steam rate/actual steam rate.

$$\text{Ideal steam rate} = 3413/\text{ideal heat drop}$$
$$= 3413/(1340.6 - 885.06) = w_i$$
$$w_i = 7.48 \text{ lb steam per kwhr}$$
$$7.48/10 = 0.748, \text{ or } 74.8 \text{ percent}$$

c. Ideal steam rate is 7.48 lb steam per kwhr. Ideal heat drop is $H_1 - H_2$ on the Mollier diagram.

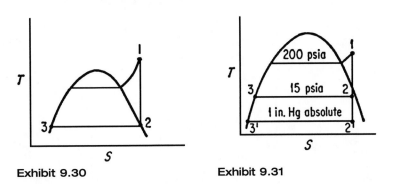

Exhibit 9.30   Exhibit 9.31

9.31 See Exhibit 9.31. Rankine efficiency is $(H_1 - H_2)/(H_1 - H_3)$. At 15 psia backpressure,

$$(1267.9 - 1062)/(1267.9 - 181.04) = 0.1892, \text{ or } 18.92 \text{ percent}.$$

At 1 in. Hg backpressure:

$$(1267.9 - 874)/(1267.9 - 47.06) = 0.3226, \text{ or } 32.26 \text{ percent}$$

Percent increase is

$$(32.26 - 18.92)/(18.92) = 0.70.51, \text{ or } 70.51 \text{ percent}.$$

9.32 Use steam tables and Mollier diagram to determine values of enthalpy. Refer also to Exhibit 9.32.

$$H_{2'} = H_2 + [0.15(H_1 - H_2)]$$
$$= 889.1 + [0.15(1240 - 889.1]$$
$$H_{2'} = 889.1 + 52.64$$
$$= 941.74. \text{ Btu per lb}$$

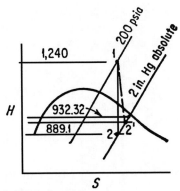

Exhibit 9.32

Heat loss by radiation = 0.01(1240 − 889.1) = 9.42 Btu per lb

$$H_{2'} - H_r = 941.74 - 9.42 = 932.32 \text{ Btu per lb}$$

Refer to the Mollier diagram. The point of intersect between enthalpy of 932.32 Btu per lb and 2 in. Hg abs pressure line shows 18.2 percent wetness. Therefore, quality is 100 − 18.2 = 81.8 percent.

9.33 Refer to Exhibit 9.33. From steam tables and/or Mollier diagram the following state points may be determined:

$$H_1 = 1290.9 \text{ Btu}$$
$$S_1 = 1.5334$$
$$H_2 = 823.5 \text{ Btu}$$
$$S_2 = 1.5334$$
$$x_2 = 0.74$$

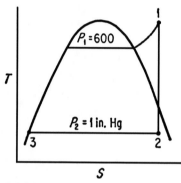

Exhibit 9.33

Ideal heat drop is ideal work = $H_1 - H_2$ = 1290.9 − 823.5
= 467.4 Btu per lb.

Ideal Rankine steam rate is

$$\frac{3413}{H_1 - H_2} = \frac{3413}{467.4} = 7.3 \text{ lb steam per kwhr.}$$

Actual work energy conversion = 467.4 × 0.73 (engine eff)
= 342 Btu per lb

a. Turbine steam rate = 3413/342 = 10 lb per kwhr
b. Pounds steam per hr to throttle = 5000 kw × 10 = 50,000 lb per hr
c. Turbine heat rate = input, Btu/output, kwhr

Input = (steam to throttle + steam to auxiliaries)($H_1 - H_3$)
Input = (50,000 + 4000)(1290.9 − 127.9)

$$\text{Turbine heat rate } \frac{\text{input}}{5000 \text{ kwhr}} = 12,560 \text{ Btu per kwhr}$$

Note enthalpy equal to 127.9 Btu per lb is that of saturated liquid at the turbine backpressure.

d. Station heat rate = $\dfrac{\text{turbine heat rate}}{\text{boiler efficiency}}$

   Turbine heat rate = $\dfrac{\text{input}}{5000 - 100 \text{ (for aux.)}}$ = 12,800 Btu per kwhr

   Station heat rate = $\dfrac{12,800}{0.80}$ = 16,000 Btu per kwhr

e. Station efficiency = $\dfrac{3413}{\text{station heat rate}} = \dfrac{3413}{16,000}$ = 0.214, or 21.4 percent

9.34 We are looking for the ratio of total available energy to Btu per kwhr. Refer to Exhibits 9.34a and b. Use the Mollier diagram and steam tables freely. We can then set up the following values:

$$H_1 = 1362.1 \text{ Btu} \qquad S_2 = 1.6393 \qquad H_x = 1285 \text{ Btu}$$
$$S_1 = 1.6393 \qquad H_2 = 1225 \text{ Btu} \qquad P_x = 200 \text{ psia}$$

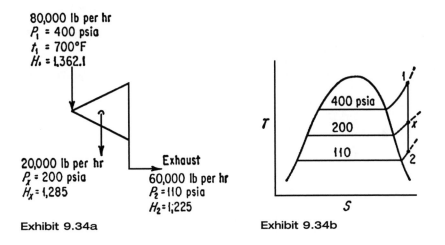

Exhibit 9.34a   Exhibit 9.34b

Total available energy for extraction:

$$80,000 \times (H_1 - H_x) = 80,000 \times (1362 - 1285)$$
$$= 6,160,000 \text{ Btu per hr}$$

Total available energy for exhaust:

$$(80,000 - 20,000)(H_x - H_2) = 60,000(1285 - 1225)$$
$$= 3,600,000 \text{ Btu per hr}$$

kw output $\dfrac{9,760,000}{3413}$ = 2860 kw

Because turbine efficiency is not involved, the 2860-kw answer is entirely theoretical. The power transfer to the generator shaft would have to take into account the mechanical efficiency of the turbine as a whole.

9.35   For the purposes of this problem the curve of total steam consumption (Willans's line) is practically a straight line. This is a characteristic of such curves for practically all types of steam turbines when operating without overloading. If Willans's line is extended to intercept the Y axis (total steam per hour), this intercept represents the steam required to operate the turbine when delivering no power. It is the amount needed to overcome the friction of the turbine and the windage, that required for driving the governor, oil pumps, etc., and that for meeting the losses due to turbulence, leakage, and radiation under no-load conditions.

The steam rate may be determined from the simple expression

$$\frac{50}{L} + 9.5 = \frac{F}{L} = \frac{50 + (1000 - 50)/100 \times L}{L}$$

where F is expressed in terms of per 1000 lb of steam per hr and from the conditions of the problem. Then

Load Fraction	Load (L), mw	Steam Rate, lb per kwhr
$\dfrac{1}{4}$	$100 \times \dfrac{1}{4} = 25$	$\dfrac{50}{25} + 9.5 = 11.5$
$\dfrac{2}{4}$	$100 \times \dfrac{2}{4} = 50$	$\dfrac{50}{50} + 9.5 = 10.5$
$\dfrac{3}{4}$	$100 \times \dfrac{3}{4} = 75$	$\dfrac{50}{75} + 9.5 = 10.17$
$\dfrac{4}{4}$	$100 \times \dfrac{4}{4} = 100$	$\dfrac{50}{100} + 9.5 = 10.00$

CHAPTER

Internal-Combustion Engines and Cycles

OUTLINE

PROBLEMS

10.1 A gas engine is supplied with 1100 cfh of gas at 80°F and 5 in. WG. Barometer reads 30.2 in. Hg. Heating value of gas is 540 Btu per cu ft at 60°F and 30 in. Hg. Engine develops 52 hp. Calculate the thermal efficiency of the engine.

10.2 Calculate the temperature and pressure at the beginning of the expansion stroke in a gasoline engine, assuming no heat losses and combustion completed. The air-to-fuel ratio is 15, compression ratio is 6.2, the compression curve exponent is 1.3, and the pressure at the beginning of the compression stroke is 13 psia with a temperature of 150°F. Assume the heating value of the gasoline to be 20,000 Btu per lb and the specific heat during expansion to be 0.3. Assume gases to be 1500°F and 17 psia when remaining in the clearance volume from the previous explosion cycle.

10.3 Calculate the pressure and temperature at 90 percent of the completion of the compression stroke in a four-cycle diesel engine having a compression ratio of 14. The air temperature at the beginning of the stroke is 150°F and the pressure is 13.7 psia.

10.4 The volume in the clearance space of a 6- by 10-in. Otto gas engine is 0.06 cu ft. Find the ideal thermal efficiency (ITE) of the engine on the air standard basis, if the exponent of the expansion and compression lines is 1.35.

10.5 A two-stroke-cycle internal-combustion engine operating at 3000 rpm has cylinders $3^{1}/_{2}$ in. in diameter with a 5-in. stroke. Intel valves are $1^{1}/_{4}$ in. in diameter, open to a mean height of $^{3}/_{2}$ in., and are open for 120 degrees of engine rotation. The air-fuel mixture is at a pressure of about 14.7 psia and temperature of 140°F. If a discharge coefficient of 0.70 is assumed for the inlet ports, what pressure differential is required to transfer a charge, equal to the engine displacement, from the inlet manifold to the cylinder?

10.6 Calculate from the following data the number of pounds of fuel used by an automobile engine: air temperature 70°F, barometer 30.2 in. Hg, air entering 60 cfm, measured gasoline 30 pints per hr, specific gravity of gasoline 0.735.

10.7 A 2-stroke-cycle engine has a 4- by 6-in. cylinder with a connecting rod 10 in. long. The intake port is 1 in. high. Determine the mean port opening during the time required to transport the required charge of fuel and air.

10.8 An automobile engine has a rating of 250 bhp at 4500 rpm. The engine torque peaks at 2000 rpm, and the horsepower peaks at 4500 rpm. The fuel has a heating value of 20,000 Btu per lb, and the overall efficiency is 28 percent. Find the fuel consumption in pounds per hour when the engine is running at 4500 rpm.

10.9 A Carnot engine and air-standard Otto, diesel, and gas-turbine engines are each operating with a heat-addition rate of 10,000 Btu/h. The Carnot engine is operating heat reservoirs at 1200 and 300°F. The other engines have compression ratios of 8:1, and the diesel engine has an expansion ratio of 2:1. Determine the net horsepower produced by
a. Carnot engine
b. Otto engine
c. Diesel engine
d. Gas-turbine engine

SOLUTIONS

10.1 Efficiency is equal to ratio of output to input. The input is

$$\frac{1100 \text{ cfh} \times \text{actual heating value under operating conditions}}{2545}.$$

Actual heating value corrected for temperature and pressure is

$$540 \times \frac{450+80}{460+60} \times \frac{30}{30.2+5/13.6} = 550 \text{ Btu per cu ft firing}$$

$$\text{Input} = 1100 \times 550/2545 = 238 \text{ hp}$$

$$\text{Efficiency} = \frac{52}{238} = 0.219, \quad \text{or} \quad 21.9 \text{ percent}$$

Be sure in gas-combustion problems that heating value at standard conditions is corrected to actual firing conditions.

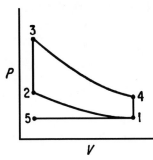

Exhibit 10.2

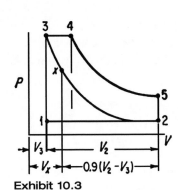

Exhibit 10.3

10.2 Assume a four-cycle engine and refer to pressure-volume diagram (Exhibit 10.2). Solve for pressure and temperature at point 3. $P_1 = 13$ psia $= P_5$. $T_1 = 460 + 150 = 610°R$. Compression ratio $= r_c = V_1/V_2 = 6.2$. Clearance volume $= V_2$. Then

$$P_2 = P_1 r_c^n = 13 \times 6.2^{1.3} = 165 \text{ psia}$$

$$T_2 = T_1 r_c^{n-1} = 610 \times 6.2^{0.3} = 1045°R, \quad \text{or} \quad 585°F.$$

This is the temperature-pressure condition of the mixture after compression and just before ignition. The addition of the heat of combustion will increase temperature and pressure. Then the heat added to this mixture, since there are no heat losses, is that given up on combustion of 1 lb of gasoline. Because this is a constant-volume process, the temperature rise $(T_3 - T_2)$ is given by

$$T_3 - T_2 = \frac{\text{Btu transferred by combustion}}{\text{wt of mixture} \times \text{sp ht}}$$

$$= \frac{20,000}{15+1} \times \frac{1}{0.3} = 4160°F.$$

From this $T_3 = 5205°R$, or $4745°F$. If we assume the validity of the perfect-gas law and V_1 is fresh ($\frac{1}{15}$) mixture

$$P_3 = P_2 \frac{T_3}{T_2} = 165 \frac{5205}{1045} = 825 \text{ psia}.$$

10.3 Refer to Exhibit 10.3. Assume n for air equal to 1.3.

$$r_c = \frac{V_2}{V_3} \quad t_2 = 150°F \quad P_2 = 13.7 \text{ psia}$$

$$P_x = P_2 \left(\frac{V_2}{V_x}\right)^n \quad T_x = T_2 \left(\frac{V_2}{V_x}\right)^{n-1}$$

Obviously, the problem is to find V_2/V_x.

$$V_x = V_2 - 0.90(V_2 - V_3) = V_2 - 0.90 \left(V_2 - \frac{V_2}{14}\right)$$

$$V_x = (2.3/14) (V_2)$$

By rearrangement

$$\frac{V_2}{V_x} = 6.08$$

$$P_x = (13.7)(6.08)^{1.3} = 142.5 \text{ psia}$$

$$T_x = (150 + 460)(6.08)^{0.3} = 1048°R, \quad \text{or} \quad 588°F$$

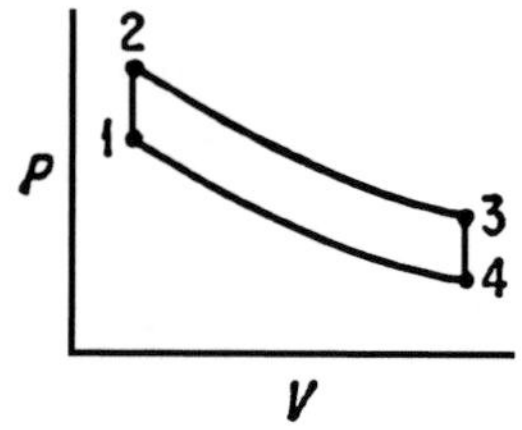

Exhibit 10.4

10.4 Refer to Exhibit 10.4. $V_1 = V_2 = 0.06$ cu ft. In the usual manner piston displacement $V_4 - V_1$ is found to be equal to 0.163 cu ft. First find V_4 by use of polytropic process and n equal to 1.35.

$$V_4 = 0.163 + 0.06 = 0.223 \text{ cu ft}$$

The ideal thermal efficiency is

$$\text{ITE} = 1 - \left(\frac{V_1}{V_4}\right)^{n-1} = 1 - \left(\frac{0.06}{0.223}\right)^{1.35-1} = 0.369, \quad \text{or} \quad 36.9 \text{ percent.}$$

10.5 The velocity of a fluid passing through an opening or aperture is

$$v = c\sqrt{2g\,\Delta H}.$$

Because pressure differential is the object of the solution, ΔH is easily converted: $\Delta H = [(P_1 - P_2)/w] \times 144$.

In this equation we know c, and g (32.2 gravitational constant). We can determine w, the air density under the conditions, and also calculate v.

$$w = \frac{29}{379} \times \frac{520}{600} \times \frac{14.7}{14.7} = 0.0662 \text{ lb per cu ft.}$$

To find velocity, we shall calculate piston displacement and then the flow rate in cubic feet per second. Then from the port area we can calculate velocity from $v = Q/A$. All calculations are based on inlet valve being open. Then

$$PD = \frac{\pi D^2 L}{4} = \frac{\pi (3.5/12)^2 (3.75/12)}{4} = 0.0209 \text{ cu ft.}$$

Note that the full stroke of 5 in. was not used, but as required by the problem the stroke for 120° of revolution $= \frac{5}{2} + \frac{5}{2}\sin 30° = 3.75$ in. For the time element involved, i.e., for the piston to move 3.75 in., the time for 1 rev. is $t = 1/(3000/60)$, or $\frac{1}{50}$ sec. This is for a total of 360°. For 120°: $t = \left(\frac{120}{360}\right)\left(\frac{1}{50}\right) = \frac{1}{150}$ sec. The flow rate Q then becomes

$$Q = 0.0209/(1/150) = 3.13 \text{ cfs.}$$

The area of the valve port is $A = 0.0085$ sq ft for a diameter of 1.25 in. When valve is in the raised position, a larger opening is available, i.e.,

$$\frac{\pi \times 1.25 \times 3/4}{144} = 0.02045 \text{ sq ft.}$$

Because the constriction occurs at the port itself, it would be more reasonable to use 0.0085 sq ft to find the velocity (higher will require a greater pressure differential). Thus, the average velocity $v = Q/A = 3.13/0.0085$, or 368 fps. Substitute the proper values in the first equation for velocity.

$$368 = 0.70 \sqrt{64.4 \times \frac{\Delta P}{0.0662} \times 144}$$

$$\Delta P = P_1 - P_2 = 1.97 \text{ psi}$$

10.6 Air density at conditions given is first determined.

$$\left(\frac{29}{379}\right)\left(\frac{520}{530}\right)(30.2/30.0) = 0.0755 \text{ lb per cu ft}$$

Weight of air drawn into the cylinders of the engine is

$$0.0755 \times 60 \times 60 = 272 \text{ lb per hr.}$$

Weight of gasoline consumed in the same period is

$$\left(\frac{30}{8}\right)\left(\frac{1}{748}\right)(62.4)(0.735) = 23 \text{ lb per hr.}$$

Pounds of air per pound of fuel: $\dfrac{272}{23} = 11.85$

10.7 The mean port opening is its area divided by height of opening. The mean port area is

$$\frac{\text{Displacement, cfs} \times 144}{\text{Charge velocity, fps}} = \text{sq in.}$$

In order to get displacement we need to know the speed at which the machine is operating and the velocity of charge. The problem does not give these data so that it must be assumed in accordance with good practice.

Assume 1800 rpm and 200 fps charge velocity. Then

$$\text{Displacement} = \frac{30 \times (\pi 4^2/4) \times 6}{1728} = 1.31 \text{ cfs}$$

where 30 is suction strokes per sec = rpm/60 = 1800/60. The mean port area is $1.31 \times \frac{144}{200} = 0.942$ sq in. Finally, the mean port opening is $0.942/1 = 0.942$ in.

10.8 Brake energy converted to heat = 250 hp × 2545 = 6.3625×10^5 Btuh

Fuel conversion = $6.3625 \times 10^5/20,000 = 31.81$ lb per hr
Actual fuel required = $31.81/0.28 = 113.6$ lb per hr at peak load
Brake hp = $(2\pi NT)/33,000$ and $T = \text{bhp} \times 33,000/2\pi N$

At peak load,

$$\text{Torque} = 250 \times 33{,}000/(2\pi \times 4500) = 292 \text{ lb-ft}$$

10.9 a. Carnot engine

$$\eta = \frac{T_1 - T_2}{T_1} = \frac{1200 - 300}{1200 + 460} = 0.5422$$
$$W = Q_1\eta = (10{,}000)(0.5422) = 5422 \text{ Btu/h}$$

Converting to horsepower

$$\text{hp} = \frac{5422 \times 788 \text{ ft} - \text{lb/Btu}}{550 \times 3600} = 2.13 \text{ hp}$$

b. Otto engine, $r = 8/1 = 8$

$$\text{Efficiency } \eta = 1 - (1/r)^{r-1} = 1 - (1/8)^{0.4}$$
$$= 0.5647 \quad \text{or} \quad 56.47 \text{ percent}$$
$$W = Q_1\eta = 10{,}000 \times 0.5647 = 5647 \text{ Btu/h}$$

Converting to hp in the same manner as above, we obtain 2.22 hp

c. Diesel engine, $r = 8$ and $r_e = 2$

$$\eta = 1 - \frac{r}{r} \frac{r_e^{-r} - r^r}{r/r_e - 1} = 1 - \frac{8}{1.4} \times \frac{2^{-1.4} - 8^{-1.4}}{8/2 - 1}$$
$$= 0.3819 \quad \text{or} \quad 38.19 \text{ percent}$$
$$W = Q_1\eta = 10{,}000 \times 0.3819$$
$$= 3819 \text{ Btu/h or the equivalent of 1.5 hp}$$

d. Gas-turbine engine, $r = 8$

$$\eta = 1 - (1/r)^{r-1/r} = 1 - (1/8)^{0.4/1.4} = 0.4480$$
$$W = Q_1\eta = 4480 \text{ Btu/h} \quad \text{or} \quad 1.76 \text{ hp}$$

Gas Turbines and Cycles

OUTLINE

PROBLEMS

11.1 An airplane supercharger consists of a turbine and a rotary compressor. It is receiving exhaust from the engine at 15 psia and 1000°F, and compressing air from 5 to 15 psia, the initial temperature being 60°F. What is the lowest efficiency it can have and still do the job?

11.2   a.   Sketch the aircraft turbojet cycle on *TS* coordinates indicating the effect of inefficiencies in all apparatus except the diffuser.

   b.   The diffuser (ram) of a turbojet engine has an efficiency of 100 percent. Estimate the temperature and pressure of air leaving the ram when it is flying at sea level at a speed of 500 mph.

   c.   What is the relationship between turbine work and compressor work in a turbojet engine?

   d.   What air flow is required to produce a thrust of 3000 lb at an air speed of 500 mph and a jet exit velocity of 1600 fps?

11.3 At a certain section of an air stream the Mach number is 2.5, the stagnation temperature is 560°R, and the static pressure is 0.5 atm. Assuming that the flow is steady, isentropic, and follows one-dimensional theory, calculate (all at the point where the Mach number is 2.5) (a) temperature, (b) stagnation pressure, (c) velocity, (d) specific volume, and (e) mass velocity.

11.4 A jet plane flies at an altitude of 5000 ft where atmospheric temperature is 40°F. An observer on the ground notes that he hears the sound of the plane exactly 3 sec after the plane has passed directly overhead. Assuming that the velocity of sound remains constant at its value corresponding to 40°F, estimate the speed of the jet plane.

11.5   Air with a specific heat ratio equal to 1.4 flows at supersonic velocity in a duct whose cross-sectional area is 2 sq ft. Static temperature in the main body of flow is estimated to be 400°R. Static pressure measurements by means of a mercury manometer indicate a manometer deflection of 10 in. of mercury below atmospheric. Pitot tube measurements indicate a manometer deflection of 90 in. of mercury above atmospheric pressure. The barometric pressure is 30 in. of mercury abs. Determine the mass flow rate of air in the duct.

11.6   A convergent-divergent nozzle operates with a constant stagnation pressure of 100 psia but under variable backpressure conditions. It is noted during start-up that when the backpressure reaches 91.8 psia, flow through the nozzle becomes critical; that is, any further reduction in backpressure does not vary the mass flow rate. As the backpressure is lowered beneath 91.8 psia, plane one-dimensional normal shock sets in so as to result in pressures at the exit plane being identical with backpressure. What is the maximum backpressure that can exist and still result in supersonic flow for the entire length of the divergent portion of the nozzle?

11.7   A stationary gas-turbine power plant is to deliver 20,000 hp to an electric generator. The plant is to have no regeneration. The maximum temperature in the turbine is set at 1540°F, and the system is designed for a maximum pressure of 60 psia. The compressor air intake is located outside the building, where the mean temperature is 60°F. For the preliminary design of the plant, determine the air intake required in cubic feet per minute.

SOLUTIONS

11.1 The turbine uses the products of combustion to generate work. For lowest efficiency,

$$\frac{\text{Minimum compressor work required}}{\text{Energy supplied to the turbine}}.$$

The minimum compressor work required may be developed through isothermal compression. Using the equation for isothermal compression for 1 lb air,

$$Q = \frac{wRT}{J} 2.3 \log \frac{V_2}{V_1}$$

and $$P_1 V_1 = P_2 V_2$$

where

$$5 \text{ psia} = P_1$$
$$15 \text{ psia} = P_2$$
$$60°F = t_1$$

$$\frac{V_2}{V_1} = \frac{P_2}{P_1}$$

$$Q = 1 \times 53.3 \times (460 + 60)/778 \times 2.3 \log \frac{15}{5} = 30 \text{ Btu.}$$

This is equivalent to $30 \times 778 = 23,400$ ft-lb. This is minimum work. The energy supplied to the turbine is equivalent to the heat drop from 15 psia and 1000°F to that assumed as 60°F. A good average c_p is 0.26.

$$c_p(t_1 - t_4) = 0.26(1000 - 60)\ 778 \equiv 190,000 \text{ ft-lb}$$
$$\text{Efficiency} = 23,400/190,000 = 0.123, \quad \text{or} \quad 12.3 \text{ percent}$$

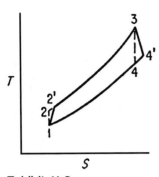

Exhibit 11.2a

11.2 a. For the effect of inefficiencies, see Exhibit 11.2a. The increases in entropy are due to internal friction in the compressor (1-2) and in the turbine during expansion (3-4). Exhibit 11.2b helps tell the story too.

b. The ramming intake increases both the temperature and the pressure at the compressor inlet. The temperature rise is the full temperature rise equivalent to the forward speed of the gas turbine.

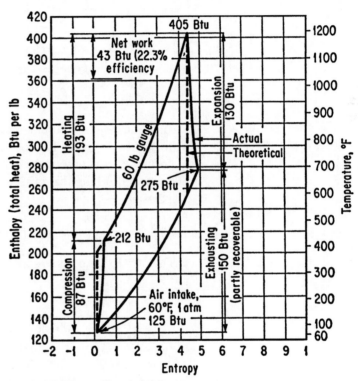

Exhibit 11.2b

Heat-entropy diagram for gas-turbine cycle resembles that for a diesel cylinder except that compression ratio is only 5:1.

$$\Delta T_r = \frac{v^2}{2gJc_p} = \frac{\left(\frac{500}{60} \times 88\right)^2}{64.4 \times 778 \times 0.26} = 41.5°\text{F}$$

Temp. leaving $= 60 + 41.5 = 101.5°\text{F}$

The pressure rise is not as efficient as the temperature rise and only a portion of the pressure rise theoretically available can be obtained. In aircraft literature, the ram efficiency is defined as follows:

$$p_1 - p_{am} = n_r q_c$$

where

p_1 = pressure at compressor inlet, psi
p_{am} = pressure of ambient, psi
n_r = ram efficiency
q_c = dynamic pressure rise, psi.

When stagnation takes place (that is, when utilizing the kinetic energy of the air stream to isentropically compress the air to a state of higher temperature (stagnation), higher pressure, and zero velocity), the pressure at stagnation may be calculated from

$$\frac{(460+101.5)}{520} = \left(\frac{p_0}{p_{am}}\right)^{(k-1)/k} = \left(\frac{p_0}{14.7}\right)^{0.283}$$
$$p_0^{0.283} = 1.08 \times 14.7^{0.283} = 1.08 \times 2.14 = 2.31$$
$$0.283 \times \log p_0 = \log 2.31$$
$$\log p_0 = 0.3636/0.283 = 1.285$$
$$p_0 = 19.3 \text{ psia.}$$

c.   In the turbojet the turbine produces work, all of which is used to drive the air compressor. The forward thrust of the plane is produced wholly from the jet action of the exhaust gases from the turbine. There is no propeller. Again, the diffuser produces part of the necessary air compression.

d.   The thrust equation may be represented as follows:

Thrust = lb air per sec/g × (jet exit velocity – plane velocity).

Then by rearrangement

$$\text{lb per sec} = \frac{3000 \text{ lb thrust} \times 32.2}{1600 - 735} = 111.8 \text{ lb per sec.}$$

11.3   The relation between Mach number M, velocity relative to some point v, fps, and acoustic velocity v_c, fps, may be expressed by

$$M = \frac{v}{v_e} = \frac{v}{\sqrt{kgRT}}.$$

The Mach number varies with position in the fluid and the compressibility effect likewise varies from point to point. If a body moves through the atmosphere, the overall compressibility effects are a function of the Mach number.
a.   For the problem at hand, Marks* gives the following formula:

$$\frac{T}{T_0} = \frac{1}{1+[(k-1)/2]M^2}.$$

Mechanical Engineers' Handbook, revised by T. Baumeister, 6th ed., pp. 11–91, McGraw-Hill Book Company, Inc., New York, 1958.

Solve for stagnation temperature T_0 by rearrangement of the above equation.

$$T_0 = T\left(1 + \frac{1.4-1}{2} \times 2.5^2\right) = 560°R$$

$T = 560/2.25 = 249°R$ the static temperature

b. See previous problem and determine stagnation pressure as follows: because process is isentropic

$$p_0 = p\left(\frac{T_0}{T}\right)^{k/(k-1)}$$

$$= 14.7 \times 0.5 \left(\frac{560}{249}\right)^{3.5} = 125.5 \text{ psia.}$$

c. $$v = Mv_c = 2.5 \times \sqrt{1.4 \times 32.2 \times 53.3 \times 249} = 1936 \text{ fps}$$

d. From the perfect-gas law

$$\bar{V} = \frac{RT}{p} = \frac{53.3 \times 249}{0.5 \times 14.7 \times 144} = 12.6 \text{ cu ft per lb}$$

e. Mass velocity may be expressed as G lb per sec-ft^2 and ρ is density in pounds per cubic foot.

$$G = \rho v = (1/12.6)(1936) = 153.3 \text{ lb per sec-ft}^2$$

For a nozzle, the stagnation condition is the state of the fluid at rest in the chamber ahead of the nozzle. For a body moving through the atmosphere, the stagnation state occurs only at the stagnation point at the nose of the body and may be computed by the formulas used in the previous and this problem.

11.4 A plane traveling slower than the speed of sound would be heard before it passed directly overhead; if at the same speed as sound (Mach 1), then it

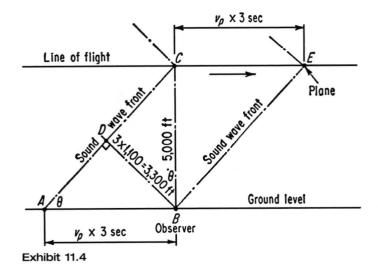

Exhibit 11.4

would be heard at the same time it passed directly overhead; if at a speed greater than sound, it would be heard after it had passed overhead. From the conditions set by the problem, make up sketch Exhibit 11.4.

$$v_c = \sqrt{1.4 \times 32.2 \times 53.3 \times (460 + 40)} = 1100 \text{ fps}$$

The speed of the plane must be greater than the acoustic velocity. When plane is at point C, wave front AC develops and sound is heard by observer 3 sec later at point B. By this time plane has traveled distance $v_p \times 3$, where v_p is velocity of plane in fps. Sound-wave fronts run parallel and are conical in shape. Likewise, the distance AB is $v_p \times 3$. Angle CAB is equal to angle CBD. Distance $CD = \sqrt{(5000^2 - 3300^2)} = 3760$ ft. From geometry $AB/BD = BC/BD$. The Mach number at which the plane is traveling is AB/BD, because they represent the distances traveled by the plane/traveled by the sound. Thus,

$$\text{Mach No. } \frac{AB}{BD}$$

Then by similar triangles

$$\frac{BC}{BD} = \frac{5000}{3760} = 1.33$$

Velocity of plane $= 1.33 \times 1100 = 1465$ fps,   or   1000 mph

11.5 Review Problem 11.3. The sonic velocity may be calculated.

$$v_c = \sqrt{1.4 \times 32.2 \times 53.3 \times 400} = 980 \text{ fps}$$

The specific volume is

$$\bar{V} = \frac{RT}{p} = \frac{53.3 \times 400}{9.85 \times 144} = 15 \text{ cu ft per lb.}$$

Absolute pressure equivalent to 10 in. Hg below atmospheric is given by

$$\left(\frac{20}{30}\right) 14.7 = 9.85 \text{ psia.}$$

The duct velocity without light correction for temperature is found to be

$$\sqrt{2 \times 32.2 \times 90 \times 1.136 \times 62.4 \times 15} = 2480 \text{ fps}$$

Mass flow rate $G = \rho v$

$$= \left(\frac{1}{15}\right)(2480)(2) = 330 \text{ lb per sec}$$

Mach No. $= 2480/1065 = 2.33$

11.6 For any convergent-divergent nozzle with a fixed value of P_1, the inlet pressure, the flow is at first subsonic at all points along the nozzle for values of P_2 slightly less than P_1; and we have the case of the ordinary venturi tube. As the exit pressure P_3 is reduced, velocity at the throat increases, and the weight discharge also increases. When P_2 reaches the critical value such that

$$\frac{P_2}{P_1} = \left(\frac{2}{k+1}\right)^{k/(k-1)} \quad \text{or} \quad \frac{P_c}{P_1} = \left(\frac{2}{k+1}\right)^{k/(k-1)}$$

and $P_c = P_1 \times 0.528$ when k is 1.4. At critical then sonic velocity exists at the throat and further reduction in P_2 fails to increase flow. Thus flow rate is a maximum when $P_c = 0.528\, P_1 = 52.8$ psia. Here the backpressure P_3 is 91.8 psia and sonic velocity (Mach 1)

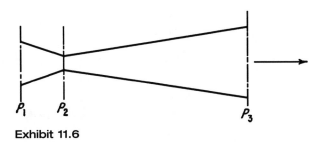

Exhibit 11.6

occurs only at the throat and no further reductions in P_2/P_1 are possible if the stream fills the passage. When the flow is entirely supersonic (Mach > 1) for the entire length of the divergent portion of the nozzle, results may be calculated by isentropic flow relations and any shock wave will occur outside the nozzle proper. Refer to Exhibit 11.6.

From isentropic gas flow tables and normal shock tables*

M	P_2/P_1	P_2 Calculated	P_3/P_2	P_3 Calculated
1.0	0.52828	52.828	1.0000	52.83
1.3	0.36092	36.092	1.8050	55.15
1.5	0.27240	27.240	2.4583	66.96
1.7	0.20259	20.259	3.2050	64.93

* Ascher H. Shapiro, *The Dynamics and Thermodynamics of Compressible Flow*, Parts I and II from Volume I, The Ronald Press Company, New York, 1958.

Thus, the maximum backpressure that can exist and still have supersonic flow appears to take place at a Mach number equal to 1.5 at a pressure of 66.96 psia. For greater accuracy examine Mach number either side of 1.5 in the same manner as above.

11.7 Assume the Brayton air standard cycle, Exhibit 11.7, and neglect pressure drop through the filter. $T_1 = (460 + 60) = 520°R$; $P_1 = 14.7$ psia $= P_4$; $P_2 = P_3 = 60$ psia; and $T_3 = (1540 + 460) = 2000°R$. Then from thermodynamics,

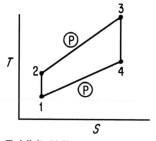

Exhibit 11.7

$$T_2 = T_1\left(\frac{P_2}{P_1}\right)^{(k-1)/k} = 520\left(\frac{60}{14.7}\right)^{0.286} = 778°R.$$

And likewise

$$T_4 = \frac{T_3}{(P_2/P_1)^{(k-1)/k}} = 1339°R$$

Heat pickup between 1 and 2 is given by

$$Q_{1-2} = c_p(T_2 - T_1) = 0.24(778 - 520) = 62 \text{ Btu per lb}$$
$$Q_{3-4} = c_p(T_3 - T_4) = 0.24(2000 - 1339) = 159 \text{ Btu per lb.}$$

Then the difference between the two states is $159 - 62 = 97$ Btu per lb. Then the air weight rate is

$$20,000 \times 42.42/97 = 8.750 \text{ lb per min.}$$

This represents a volume of $(8750 \times 53.34 \times 520)/(14.7 \times 144)$

$$= 114,000 \text{ cu ft per min.}$$

Refrigeration

PROBLEMS

12.1   Liquid ammonia expands from a tank at 180.6 psia and 90°F in an evaporating coil and is then compressed. The compressor suction is 4.3 psi below atmospheric pressure. What is the refrigerating effect per pound of ammonia?

12.2   An ammonia refrigeration unit has an 180-rpm, double-acting compressor whose cylinder has a diameter of 8 in. with a 10-in. stroke and whose clearance is 5 percent. It is desired to calculate the refrigeration capacity in tons of refrigeration at an evaporation temperature of 12°F and condensing temperature of 105°F. Assume that the 12°F vapor is saturated as it enters the compressor. Also assume that the liquid ammonia is not subcooled at the entrance to the expansion valve.

12.3   Ammonia enters the cooler of an ammonia refrigerating machine at 0°F and leaves it at 15°F. When operating at 10 tons of refrigeration (ice melting capacity) and 15 indicated horsepower at the compressor, how many gallons per minute of condensing water must be supplied for a temperature rise from 50°F at entrance to 70°F at condenser exit? Assuming the liquid ammonia to leave the condenser at 50°F, calculate the ideal coefficient of performance. What is the actual coefficient of performance?

12.4   In a test on an ammonia refrigerating machine, 30 gpm of water are collected at the condenser outlet. For the same period the water enters the condenser at 70°F and leaves at 85°F. Assume "28 gal-deg per ton of refrigeration." What is the system capacity?

12.5   (a) In an ammonia ice machine, how many pounds of ice at 24°F are produced by the evaporization of 1 lb of ammonia, the liquid ammonia entering the cooler at 60°F and leaving as a vapor at 20°F? The pressure in the expansion coils is 15 psig. Cooler efficiency is 75 percent. Water is placed in the cooler at 50°F. (b) What is the ideal COP in this case?

12.6   (a) What characteristics are necessary in a refrigerant to produce a high COP? (b) Assuming an actual COP of 3, calculate the indicated horsepower required to produce 1 TR. (c) With a temperature of 15°F in the cooler and condensing water available at 50°F, calculate the ideal COP.

12.7   An ammonia compressor works between a pressure of 40 psia, 30°F, and 100 psia.
a.   What is the final temperature if the compression is adiabatic?
b.   What work is required per pound of ammonia if the compression is adiabatic?
c.   If the compressor is water-jacketed so that the final compression temperature is 180°F, how many foot-pounds of work are required per lb of ammonia circulated?
d.   What percent saving in energy to drive the compressor is made by the use of the water jacket?
e.   How many gpm of water must be circulated through the water jacket per pound of ammonia if the temperature rise in the water jacket is 10°F?
f.   If expansion takes place in the expansion coils from liquid ammonia at 75°F to ammonia vapor at 40 psi and 25°F, how much refrigerating effect is obtained per pound of ammonia?
g.   What is the COP?
h.   For a capacity of 200 TR, what horsepower is required in each case to drive the compressor if the mechanical efficiency of the compressor is 80 percent?
i.   For a capacity of 200 TR, how many pounds of ammonia must be circulated per minute in each case?
j.   How many gpm circulating water must be used in the compressor jacket?
k.   If cooling water enters the condenser at 60°F and leaves at a temperature 40°F below that of the entering ammonia, how many gpm of circulating water are required in the condenser in each case?
l.   If the compressor clearance is 5 percent and the piston speed is 600 fpm, what is the cylinder diameter if the compressor is single-acting?
m.   If the length of stroke is 1.25 times the diameter of the cylinder and the compressor is to be belt-driven from a motor making 900 rpm and having a pulley 14 in. in diameter, what diameter pulley should be provided on the compressor?

12.8   An ice plant is freezing 1000 gal of water per hr from 65°F into ice at 20°F. Find the output of the refrigeration machine in tons, assuming 20 percent additional load due to losses in heat leakage, etc.

12.9   In a simple single-stage ammonia refrigeration system, the cooler is supplied with saturated liquid from the condenser at 180 psig, and the compressor is drawing saturated gas from the cooler at 25 psig. The compressor is a 10- by 12-in. single-cylinder, double-acting machine running at 30 rpm. If the volumetric efficiency is 75 percent, what is the refrigerating capacity of the compressor under these conditions?

12.10 In an industrial plant it is planned to insulate a large steel reservoir in a 15°F brine refrigeration system with molded cork slabs. Refrigeration costs $1 per ton. Insulation is available in 1- to 5-in. thickness with 1-in. variations at installed costs of 22.5 cents, 35 cents, 45 cents, 52 cents, and 75 cents per sq ft. What thickness of insulation should be applied to realize a minimum of 25 percent return on the cost of insulation?

12.11 A single-stage propane compressor driven by an electric motor in a refrigeration system will be used under a variety of suction conditions but at a constant speed and constant discharge pressure. Cooling water is available at 70°F and will be allowed to rise to 90°F. Minimum temperature differential in the condenser is to be 10°F. Evaporator temperature on the propane side will vary from −30 to −10°F. Saturated propane vapor will enter the suction side of the compressor. Assume propane to behave like a perfect gas. Neglect variations in volumetric efficiency and power efficiency due to change in suction conditions. Estimate the approximate maximum horsepower necessary to drive the compressor per 100 cu ft of intake volume per minute. Estimate k equal to 1.13.

12.12 Ammonia is to be liquefied by compression and cooling. Cooling water is available at 60°F, but because it is limited in quantity, it is desired that the temperature rise to 30°F in the condensing section of the cooler. A log mean overall Δt of 20°F in the condensing section is also desired. Estimate the work in kilowatt-hour required per 100 lb ammonia liquefied if the compression is adiabatic and ammonia enters as a saturated vapor at 30 psia. (a) First assume ammonia is an ideal gas and then (b) repeat the calculation without making the assumption. Assume that any superheat is removed in a separate cooler ahead of the condenser.

12.13 A Freon-12 refrigeration unit has a double-acting compressor 6 by 10 in. with 6 percent clearance, and runs at 200 rpm. What is the rated capacity of the unit in standard tons? How many tons of refrigeration will it furnish if the Freon is evaporated at −12°F and condensed at 110°F? Assume k of 1.14.

12.14 A propane refrigeration system in a processing plant supplies 80,000 Btuh of refrigeration at 20°F (temperature of the refrigerant). Cooling water may be assumed to be 80°F, condensing the propane at 100°F. Refer to Exhibit 12.14 for T-S relations.
  a. Determine conditions at exit of condenser.
  b. After expansion at constant enthalpy, −20°F propane is saturated at 25.05 psia and enthalpies are equal. What is the new enthalpy?
  c. After evaporation, vapor at −20°F is saturated at the backpressure of 25.05 psia. What is the enthalpy?
  d. Determine the refrigeration effect.
  e. Determine the propane circulation rate to maintain the refrigeration effect.
  f. Determine the ideal work of compression under adiabatic no-work conditions, $\Delta S = 0$. Follow the constant entropy line from −20°F saturated propane to 188.7 psia in the superheat region of the T-S curve.

Pressure, psia

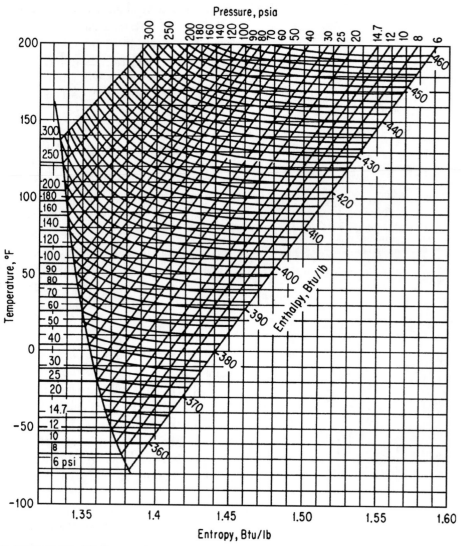

Exhibit 12.14 *TS* diagram for propane

    g. Determine the actual work of compression, assuming compressor efficiency of 75 percent.

    h. Determine the brake horsepower required to compress the propane circulated.

    i. Determine the heat to be removed by the cooling water.

    j. Determine the amount of cooling water required for this heat removal.

12.15 An ammonia refrigeration unit has a 180-r/min double-acting compressor whose cylinder has a diameter of 8 in. with a 10-in. stroke whose clearance is 5 percent.

    a. Sketch the *TS* diagram of the cycle.

    b. Sketch the mechanical hookup of the cycle.

    c. Calculate the refrigeration capacity of the unit in tons of refrigeration.

Data and Assumptions:

Evaporation temperature, 12°F
Condensing temperature, 105°F
Vapor is saturated as it enters the compressor suction.

Liquid ammonia is not subcooled at the entrance to the expansion valve.

	At 12°F	At 105°F
P, lb/in^2 abs	40.31	228.9
v_f, ft^3/lb	0.02451	0.02769
v_g, ft^3/lb	6.996	1.313
h_f, Btu/lb	56.0	161.1
h_{fg}, Btu/lb	559.5	472.3
h_g, Btu/lb	615.5	633.4
S_f, entropy units	0.1254	0.3269
S_{fg}, entropy units	1.1864	0.8366
S_g, entropy units	1.3118	1.1635

SOLUTIONS

12.1 Pressure of ammonia leaving coil and entering compressor suction is $14.7 - 4.3 = 10.4$ psia. This assumes no pressure drop in the suction line. From ammonia tables for dry and saturated (not superheated) ammonia, the enthalpy of saturated ammonia vapor at 10.4 psia is 597.6 Btu per lb. The enthalpy of saturated liquid at 180.6 psia is 143.5 Btu per lb. Note also that the temperature of 90°F given is also the saturation temperature at the pressure. Then, the refrigerating effect per pound circulated is

$$597.6 - 143.5 = 454.1 \text{ Btu per lb circulated.}$$

12.2 The following table lists the various thermodynamic data for both the evaporation and condensing temperatures:

Condition	12°F	105°F
Pressure, psia	40.31	228.9
Sp vol liquid, cu ft per lb	0.02451	0.02769
Sp vol vapor, cu ft per lb	6.996	1.313
Enthalpy of liquid, Btu per lb	56.0	161.1
Enthalpy of vaporization, Btu per lb	559.5	472.3
Enthalpy of vapor, Btu per lb	615.5	633.4
Entropy units liquid	0.1254	0.3269
Entropy vaporization units	1.1864	0.8366
Entropy units vapor	1.3118	1.1635

Refer to Exhibit 12.2a for TS diagram and Exhibit 12.2b for diagram of refrigeration-cycle mechanical presentation. Tons of refrigeration TR is given by

$$\text{TR} = \frac{VDV_e(H_1 - H_4)}{200 \times \text{sp vol sat vapor}}.$$

Refrigerating effect with no liquid subcooling is $H_1 - H_3$ or $H_1 - H_4$ with only slight liquid flashing. VD is volumetric displacement of compressor

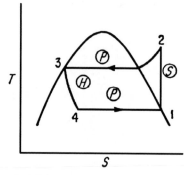

Exhibit 12.2a

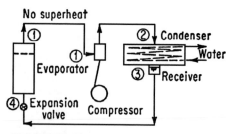

Exhibit 12.2b

at 100 percent volumetric efficiency $V_e = 1$. If we neglect the area of the piston rod to simplify calculations,

$$VD = \text{rpm} \times 2 \times \frac{\pi d^2 L}{4 \times 1728} = 180 \times 2 \times \frac{\pi \times 8^2 \times 10}{4 \times 1728} = 105 \text{ cfm}.$$

The equation for volumetric efficiency with c as clearance (decimal) and the familiar polytropic exponent n is

$$V_e = 1 + c - c \left(\frac{P_2}{P_1} \right)^{1/n}.$$

Thus, the volumetric efficiency of this machine is

$$V_e = 1 + 0.05 - 0.05 \left(\frac{228.9}{40.31} \right)^{1/1.3} = 1.05 - 0.19 = 0.86$$

$$\text{TR} = \frac{105 \times 0.86 \times (615.5 - 161.1)}{200 \times 6.996} = 29.5 \text{ tons}$$

12.3 Heat pickup in evaporator is

$$10 \times 200 = 2000 \text{ Btu per min.}$$

Heat added in compressor cylinder is 15 hp × 2545/60 = 634 Btu per min. Thus the total heat pickup is the sum of the two items 2000 + 634 = 2634 Btu per min, which is to be absorbed by the condensing water through the 20°F range. The flow of condensing water is

$$\text{gpm} = \frac{2634}{20 \times 8.33} = 15.9 \text{ gpm.}$$

The ideal coefficient of performance (COP) is

$$(460 - 0)/(70 - 0) = 6.57.$$

The actual COP is

$$\frac{12,000 \text{ Btu per hr per TR} \times 10}{15 \times 2545} = 3.14.$$

12.4 The figure of 28 gal-deg is a common factor used to test refrigeration systems with water-cooled condensers. This means that 28 gpm will be raised 1° or 1 gpm will be raised 28° to remove the equivalent of 1 ton of refrigeration from the system. This takes into account both sensible and superheat added to the refrigerant when passing through the compressor. Now for the problem at hand,

$$\frac{30 \times (85 - 70)}{28} = 16 \, TR.$$

12.5 Assume no liquid subcooling in condenser and liquid expands through float valve directly into the cooler. There is no pressure drop assumed to take place in the piping. Temperature leaving condenser and entering float valve is 60°F, and its corresponding saturation pressure is 107.6 psia. The pressure in the cooler after expansion is 15 + 14.7 = 29.7 psia, say 30 psia. The temperature therein is, therefore, 0°F at saturation.

Suction temperature is 20°F. Suction pressure is the same as in the coils, assuming no pressure drop, and is equal to 30 psia. The compressor discharge pressure is the same as the condenser, or 107.6 psia. At isentropic compression temperature is 182°F.

Heat (enthalpy) of liquid before expansion at 60°F is 109.4 Btu per lb. Enthalpy of saturated vapor at compressor suction (no superheat) is 623.5 Btu per lb. Enthalpy of compressor discharge vapor is 702 Btu per lb.

a. Ice formed per lb ammonia:

Heat absorbed by ammonia is 623.5 − 109.4 = 514.1 Btu per lb. Heat removed from water per lb is

From 50 to 32°F	18 Btu
During freezing	144 Btu
Ice[*] cooling 32 to 24°F	4 Btu
Grand total	166 Btu per lb

[*]1 lb × 0.5 sp ht × (32 − 24) = 4 Btu.

Ice formed = 514.1 × 0.75/166 = 2.2 lb ice per lb ammonia

b. COP.

$$\text{Actual COP} = \frac{514.1}{702 - 623.5} = 6.55$$
$$\text{Ideal COP} = (460 - 0)/(60 - 0) = 7.7$$

12.6 a. COP is an indication of *refrigerating effect* in the evaporator for the work of compression. Therefore, the characteristics of a refrigerant must give high refrigerating effect for the low work of compression. Thus, the necessary characteristics are

1. Large latent heat of vaporization at moderately low pressure.

2. High critical point to prevent excessive power consumption.

3. Evaporation pressures above atmospheric to avoid leakage into the system and thus keep head pressures low to reduce power consumption and increase refrigerating effect.

4. Boiling point at atmospheric pressure lower than the lowest temperature to be produced under ordinary conditions and thus keep head pressures low. High head pressures cause excessive power consumption.

5. Low density so as to reduce pressure required to raise the head of liquid when condensing unit and evaporator are not at same level.

6. Low piston displacement to reduce friction loss and power.

7. Low viscosity to keep pressure drop small and thus power requirements low.

b. Indicated horsepower is, by dimensional analysis,

$$\text{ihp per TR} = (\text{Btu per hr per TR})$$
$$\times (\text{hr per day})/(\text{ihp} \times 2545 \times \text{hr per day})$$
$$= 12{,}000 \times 24/(\text{ihp} \times 2545 \times 24) = 3$$
$$\text{ihp} = 12{,}000 \div (3 \times 2545) = 1.58 \text{ ihp.}$$

c. Ideal COP is

$$(\text{evap. temp.})/(\text{water temp.} - \text{evap. temp.})$$
$$= (460 + 15)/(50 - 15) = 13.5.$$

12.7 With the appropriate charts and thermodynamic tables for other refrigerants, the solution to this problem could apply to Freon-12, Freon-22, methyl chloride, sulphur dioxide, etc. Refer to standard handbooks for charts and tables. See also Exhibit 12.7.

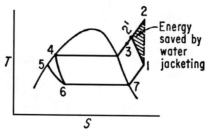

Exhibit 12.7

a. Final temperature is 240°F.
b. Adiabatic compression work is

$$H_2 - H_1 = 731 - 626 = 105 \text{ Btu per lb.}$$

c. Polytropic compression work when water jacketed is $H_{2'} - H_1$

$$693 - 626 = 67 \text{ Btu per lb.}$$

d. Percent saving polytropic over adiabatic compression:

$$\frac{105 - 67.0}{105} \times 100 = 36.2 \text{ percent}$$

e. Gallons circulated:

$$(Q_1 - Q_2)\left(T_1 + \frac{T_{2'} - T_1}{2}\right)$$

$$= (1.335 - 1.280)\left(490 + \frac{640 - 490}{2}\right) \text{ Btu per lb}$$

31.1 Btu per lb, or 31.1/10 = 3.11 lb water

$$\text{gal} = \frac{3.11}{8.33} = 0.374 \text{ gal jacket water per lb ammonia}$$

f. Refrigerating effect per pound of ammonia:

$H_4 = 623$ Btu per lb enthalpy of liquid = 127 Btu per lb
623 − 127 = 496 Btu per lb ammonia

g. COP

$$\text{COP adiabatic} = \frac{496}{105} = 4.72$$

$$\text{COP polytropic} = \frac{496}{67.0} = 7.40$$

h. $$\text{Adiabatic hp} = \frac{4.71}{472} \times 200 \times \frac{1}{0.80} = 250 \text{ hp}$$

$$\text{Polytropic hp} = \frac{4.71}{7.40} \times 200 \times \frac{1}{0.80} = 159.5 \text{ hp}$$

i. Ammonia circulated:

$$\frac{200 \text{ TR} \times 200 \text{ Btu per min per TR}}{496} = 80.8 \text{ lb ammonia}$$

j. Compressor jacket cooling water:

$$80.8 \times 0.374 = 30.2 \text{ gpm}$$

k. Adiabatic cooling water: final temperature is 240°F. Heat removed from condenser per minute is

$$200 \times 200 + (105 \times 80.8) = 48{,}490 \text{ Btu per min.}$$

With temperature of exit water at 200°F, we can set up the following heat balance:

$$8.33 \times \text{gpm } (200 - 60) = 48{,}490$$
$$\text{gpm} = 48{,}490/(8.33 \times 140) = 41.5 \text{ gpm}$$

Polytropic cooling water:

$$200 \times 200 + (67 \times 80.8) = 45{,}200 \text{ Btu per min}$$
$$\text{gpm} = 45{,}200/(8.33 \times 80) = 67.8 \text{ gpm}$$

l. Cylinder diameter. We must first determine volumetric efficiency of compressor and then the piston displacement. Assume compression exponent of 1.3.

$$E_v = 1 + 0.05 - 0.05 \left(\frac{190}{40} \right)^{1/1.3} = 1.05 - (0.05 \times 3.32) = 0.884$$

The volumetric efficiency is, therefore, 88.4 percent. Then the piston displacement is 80.8/0.884 = 91.5 lb per min. This is equivalent to 7.5 cu ft per lb × 91.5 = 686 cfm. Cylinder area is simply $\frac{686}{600} = 1.1473$ sq ft. Now

$$0.7854d^2 = 1.143 \times 144 \quad d^2 = 210 \quad d = 14.5 \text{ in.}$$

m. Pulley diameter:

$$\text{Length of stroke} = \frac{14.5 \times 1.25}{12} = 1.51 \text{ ft}$$
$$\text{Number of strokes} = 600/1.51 = 398 \text{ strokes per min}$$
$$\frac{398 \text{ strokes per min}}{2 \text{ strokes per rev}} = 199 \text{ rpm for compressor pulley}$$

Then

$$\pi \times 14 \times 900, \text{ for motor pulley}$$

$$\pi \times \text{diameter compressor pulley} \times 198$$

$$\text{Diameter of compressor pulley} = \frac{14 \times 900}{198} = 63.2 \text{ in.}$$

12.8

Heat of liquid water at 65°F	33 Btu per lb
Heat of fusion	144 Btu per lb water
Subcooling solid ice (32 − 20) 0.5	6 Btu per lb ice
Total	183 Btu per lb of water (ice)

Convert gallons of water to pounds: $1000 \times 8.33 = 8330$ lb.

Finally,

$$\frac{8333 \times 183 \times 1.20}{12,000} = 152.5 \text{ tons refrigeration.}$$

12.9 Volume of ammonia pumped is

$$\frac{0.785 \times 10^2}{144} \times \frac{12}{12} \times 2 \times 300 \times 0.75 = 245 \text{ cfm.}$$

Ammonia vapor at 25 psig and saturated is 7.0 cu ft per lb and the weight of ammonia is simply $245/7.0 = 35$ lb. From ammonia tables the enthalpy of saturated vapor at 25 psig is 615 Btu per lb while the enthalpy of the liquid at 180 psig and saturated is 149 Btu per lb. The difference of refrigerating effect is 466 Btu per lb. And the capacity of the system is

$$\frac{466 \times 35}{200} = 81.5 \text{ tons of refrigeration.}$$

12.10 Set up the following table:

Thickness of Insulation, in.	1	2	3	4	5
Installed cost per sq ft	$0.225	$0.35	$0.45	$0.52	$0.75
25 percent return per sq ft	$0.056	$0.0875	$0.1125	$0.13	$0.188

Assume steel reservoir $\frac{1}{2}$ in. thick, 50°F room temperature, 24 hr per day, and 365 days per year operation. With no insulation, heat gain in steel shell is

$$Q = UA\Delta t = 15 \times 1 \times (50 - 15) = 2100 \text{ Btu/(sq ft)(hr).}$$

Cost of heat loss with no insulation is $2100/12,000 \times \$1 = \$0.175$. Conductance with 1-in. cork insulation, assuming no surface coefficients, is

$$Q = 0.30 \times 1 \times (50 - 15) = 10.5 \text{ Btu/(sq ft)(hr)}$$

$10.5/12,000 \times \$1 = \$0.000875$. Saving $= 0.175 - 0.000875 = \$0.174125$ per sq ft. Because $0.174125 is over $0.056, use 1-in. thickness as minimum. Note that many factors were assumed to speed up development of the answer. This is acceptable if examiners do not require complete calculations.

12.11 Try

$$T_2 = 100°F \left(\frac{P_2}{P_1}\right)^{0.115} = \frac{560}{430} = 1.3$$

$$P_1 = 20.2 \quad \frac{P_2}{20.2} = 9.8 \quad P_2 = 198$$

$$T_1 = -30°F$$

Assumed T_2 of 100°F is slightly lower, but for all practical purposes call $P_2 = 198$. For

$$P_1 = 45.5 \qquad T_2 = (198/45.5)^{0.115} \times 470 = 560°R$$
$$T_1 = 10°F \qquad \text{This is acceptable.}$$
$$P_2 = 198$$

For the maximum Δt condition

$$W_1 = \frac{1.13 \times 20.2 \times 144 \times 100}{0.13 \times 33,000} \times 0.3 = -22.8 \text{ hp per 100 cu ft suction.}$$

For the minimum Δt condition

$$W_2 = \frac{1.13 \times 45.5 \times 144 \times 100}{0.13 \times 33,000} \left(1 - \frac{560}{470}\right) = -32.5 \text{ hp per 100 cu ft suction.}$$

12.12 Saturated ammonia vapor at 30 psia has a temperature of 0°F. Determine the temperature of saturated ammonia vapor entering condenser.

$$\log \text{ mean } \Delta t = 20 = \frac{30}{2.3\log[(t-60)/(t-90)]}$$

From which t is found to be 98.5°F. The vapor pressure of ammonia at 98.5°F is 206 psia from ammonia tables.

a.   Ideal gas adiabatic case power calculated

$$W = \frac{kRT_1}{k-1}\left[1 - \left(\frac{P_2}{P_1}\right)^{(k-1)/k}\right] = \frac{1.31 \times 10,987 \times 558.5}{0.31 \times 3415}\left[1 - \left(\frac{206}{30}\right)^{0.236}\right]$$

$$W = -0.656 \text{ kwhr per lb mol}$$

For 100 lb ammonia, $W_{theo} = -0.656 \times \left(\frac{100}{17}\right) = -3.84 \text{ kwhr}$

b.   Nonideal using enthalpy chart

H_1 at 30 psia and 0°F = 611.0 Btu per lb
H_2 at 206 psia (constant temperature) = 735.0 Btu per lb

$$W = -\Delta H = -124 \times 100 \times \frac{1}{3415} = -3.63 \text{ kwhr}$$

12.13 A standard ton of refrigeration is based on an evaporator temperature of 5°F and a discharge temperature of 110°F.

At 5°F: $P_1 = 26.51$ psia $H_1 = 78.79$ Btu per lb (sat)

At 14°F: $P_1 = 26.51$ psia $\overline{V}_1 = 1.52$ cf per lb

At 86°F: $P_2 = 107.9$ psia $H_2 = 27.72$ Btu per lb (sat liq)

$H_1 - H_2 = 53.7$ Btu per lb $H_2 = 25.1$ Btu per lb at 77°F (sat liq cooled)

Cylinder displacement = $(2 \times 0.785 \times 36 \times 10 \times 200)/1728 = 65.5$ cfm

Actual displacement = $65.5[1 + 0.06 - 0.06(107.9/26.51)^{0.88}] = 56$ cfm

$56/1.52 \times 53.7 = 1976$ Btu per min. This is equivalent to $1976/200 = 9.88$ standard tons of refrigeration. It would be safe to say 10.

At an evaporator temperature of −12°F

At − 12°F: $P_1 = 18.37$ psia $H_1 = 76.81$ Btu per lb (sat)

At 3°F: $P_1 = 18.37$ psia $\overline{V}_1 = 2.0$ cf per lb

At 110°F: $P_2 = 150.7$ psia $H_2 = 33.65$ Btu per lb (sat liq)

$H_1 - H_2 = 76.81 - 33.65 = 43.16$ Btu per lb

Actual displacement = $65.5[1 + 0.06 - 0.06(150.7/18.37)^{0.88}] = 44.3$ cfm
$44.3/2.0 \times 43.16/200 = 4.75$ tons of refrigeration

12.14 a. Because 100°F liquid propane is saturated at 188.7 psia, enthalpy $h_1 = 264.6$ Btu per lb from propane tables.
    b. $h_1 = h_2 = 264.6$ Btu per lb. No heat removed or work performed.
    c. $h_3 = 371.5$ Btu per lb from propane tables.
    d. $h_3 - h_2 = 371.5 - 264.6 = 106.9$ Btu per lb. This is the evaporation load.

    e. $80,000/106.9 = 748$ lb per hr.

    f. $h_5 = 413$ Btu per lb is enthalpy at the end of ideal compression.

       Ideal work = $413 - 371.5 = 41.5$ Btu per lb

    g. The actual work of compression is $41.5/0.75 = 55.3$ Btu per lb.
    h. The propane circulation rate is 748 lb per hr, and thus

                $(55.3 \times 748)/2545 = 16.25$ brake up.

    i. Conditions at compressor discharge:

       Enthalpy = enthalpy at saturation on entry to compressor suction plus work of compression

       Thus, $371.5 + 55.3 = 426.8$ Btu per lb.

This corresponds to a temperature of 151°F if no heat is removed by compressor jacket cooling. Then the heat to be removed is 426.8 − 264.6 = 162.2 Btu per lb. Finally, heat to be removed is 426.8 − 748 = 121,326 Btuh.

j. $Q = Wc_p\Delta t$, and by rearrangement, $W = Q/(c_p\,\Delta t) = 121{,}326\ (1 \times 10)$ or 12,133 lb per hr of water. Flow in gpm is 12,133/500 = 24.2 gpm.

12.15   a.   See Exhibit 12.16a.

b.   See Exhibit 12.16b

c.   Calculation of unit capacity:

$$\text{Refrigeration effect} = Q_A = h_1 - h_4 = h_1 - h_3$$

$$\text{Tons of refrigeration} = N = \frac{V_{D\eta}V(h_1 - h_4)}{200V_1}$$

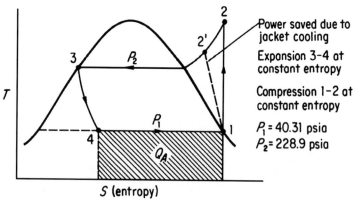

Exhibit 12.16a

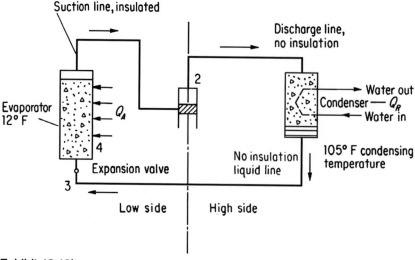

Exhibit 12.16b

where VD = Compressor displacement at 100 percent volumetric efficiency.

For a double-acting compressor and neglecting the area of the piston rod,

$$\frac{V_D = (2)(\text{r/min})(\pi d^2 L)}{4 \times 1728} = \frac{(2)(180)(\pi^2 \times 10)}{4 \times 1728} = 105 \, \text{ft}^3/\text{min}$$

$$\eta V = \text{volumetric efficiency} = 1 + c - c \left(\frac{P_2}{P_1} \right)^{1/n}$$

where

c = percent clearance as a decimal = 0.05
n = polytropic exponent = 1.3 for ammonia vapor

$$\eta V = 1.05 - 0.05 \left(\frac{228.9}{40.31} \right)^{1/1.3} = 1.05 - 0.19 = 0.86$$

h_1 = enthalpy of saturated vapor at 12°F = 615.5 Btu/lb
h_4 = enthalpy of saturated liquid at 105°F = 161.1 Btu/lb
v_1 = specific volume of saturated vapor = 6.996 ft³/lb

1 ton refrigeration = 200 Btu/min heat removal

$$N = \frac{(105)(0.86)(615.5 - 161.1)}{(200)(6.996)} = 29.4 \, \text{tons}$$

CHAPTER 13

Heating, Ventilating, and Air Conditioning

OUTLINE

PROBLEMS

13.1 The following data were obtained by a heating engineer during a survey of a client's office in New York City. The only heat loss is through one wall facing north. The exposed wall measures 20 by 10 ft. The wall has one single-pane, $\frac{1}{8}$-in. thick window, 20 sq ft in area. The wall is constructed of 4-in. face brick, 4-in. common brick, 1-in. air space, $\frac{3}{4}$-in. wood fiber insulating board (density 15 lb per cu ft), $\frac{3}{8}$-in. gypsum lath on 2- by 4-in. studs, and $\frac{1}{2}$-in. plaster. The outside design temperature is 0°F and the inside temperature is to be 70°F. Infiltration is assumed to be 10 cfm.

   a. Calculate the room heat loss.
   b. How many pounds of saturated steam at 2 psig are required to maintain temperature of 70°F?
   c. How much air at 100°F is required to heat the room?
   d. What is the temperature in the $3\frac{5}{8}$-in. air space (4-in. stud space)?

13.2 A pitched roof construction of asbestos shingles on roofing felt plus wood sheathing has an overall coefficient of 0.01 Btu/(hr)(sq ft)(°F). The roof area is 1000 sq ft. The ceiling, with an area of 600 sq ft, is made up of $\frac{3}{4}$ in. of gypsum lath and plaster having a conductance of 1.80 Btu/(hr)(sq ft)(°F) for thickness stated, plus 2 in. of mineral wool having a conductivity of 0.27 Btu/(hr)(sq ft)(°F) for 1-in. thickness. Inside and outside film conductances are 1.65 and 6.0, respectively. Outdoor temperature is plus 10°F. Indoor air is at 70°F and 20 percent relative humidity. Attic space is not ventilated. Will condensation form on the attic side of the wood sheathing? Support your answer by numerical calculations.

13.3 How many pounds of moisture must be added per hour to air entering a building at 32°F and 60 percent relative humidity to produce an inside relative humidity of 30 percent at 70°F? Building volume is 500,000 cu ft and there are three air changes per hour. Take specific volume of mixture of air and water vapor as 13.8.

13.4 Find the season fuel cost of heating a building whose heat loss calculated on the basis of 70 and 0°F is 240,000. Btu per hr, for a season of 7000 degree days using coal at $15 per ton and 12,000 Btu per lb, burned with an efficiency of 60 percent.

13.5 It is frequently stated that gases and vapors heavier than air should be exhausted at or near floor level and those lighter than air should be exhausted at or near the ceiling. Experience indicates that it is important to consider the distribution, concentration, and temperature of the gas or vapor in the air before drawing any conclusions as to the relative merits of which system to use. This is especially true of general ventilation where the forces influencing the rate of gas or vapor diffusion are complex or uncertain. In the situation that follows, these considerations are to be taken into account.

A processing room 20 ft wide by 40 ft long by 15 ft high is to be provided with general ventilation, also known as dilution ventilation, at the rate of 15 air changes per hour. The room contains 10,000 ppm of a vapor having a specific gravity (compared with air) of 5. The vapor is well mixed with room air because of drafts, plant activity, and convection currents within the room.

a. Determine the effective specific gravity of the room air. Discuss the implications and room effects.
b. Determine the cubic feet per minute of room air to be exhausted.
c. At what level or levels should exhaust points be located, at floor, midway, or at ceiling?
d. Determine the concentration of vapor in the room air within the exhaust system.
e. Estimate fan brake horsepower if air hp requirements are given by

$$\text{Air hp} = \frac{0.16 \times Ap \times \text{ft}^3/\text{min}}{1000}.$$

Assume a reasonable fan efficiency and a system duct resistance from most remote exhaust intake point to outdoor discharge to be 1.5 in. WG.
f. What would be the makeup air heating load if design conditions were +10°F outdoors and air leaving the duct heater is to be 80°F?

13.6 Ethyl ether is being evaporated from a product in a room-temperature drying hood operating at atmospheric pressure. Solvent vapor must be removed at the rate of 46 lb/h and the solvent vapor concentration must be maintained below the 1.8 percent lower explosive limit (LEL). Calculate the air volume that must be supplied to the dryer. Room air is at 75°F.

13.7 Sulfur dioxide is to be recovered from a dry gas containing (by volume) 13 percent SO_2 and 87 percent air, using a tower at atmospheric pressure with water as the solvent. By means of lead cooling coils, the temperature is maintained at 40°C throughout the operation. The raw gas, entering at the rate of 100 ft³/min at 40°C, is to be stripped of 99 percent of its SO_2 content. Data: See Exhibit 13.7. Determine the minimum amount of water that can be used.

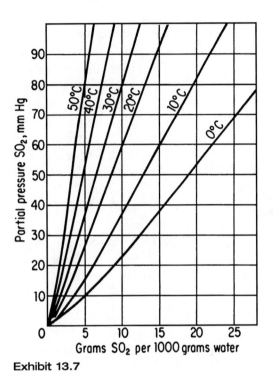

Exhibit 13.7

13.8   A 10-point pitot tube traverse for air at 70°F and normal barometric pressure of 14.7 lb/in.² abs flowing in a 12-in. circular duct gives the following data:

Traverse Point, in from Wall	Vertical Traverse, ah in WG	Horizontal Traverse, Ah in WG
3/8	0.10	0.14
1	0.11	0.18
1¾	0.19	0.22
2¾	0.27	0.30
4⅛	0.35	0.35
7⅞	0.34	0.36
9¼	0.29	0.24
10¼	0.22	0.19
10⅞	0.16	0.14
11⅝	0.13	0.11

Throat suction at inlet to the 12-in. duct is 0.44 in. WG. A subsequent throat suction measurement gives a reading of 0.28 in. WG with the inlet unchanged.
a. Determine the ft^3/min flowing through the pipe.
b. Determine the coefficient of entry C_e of the inlet. Throat suction is 0.44 in. WG.
c. Determine the quantity of air flowing through the pipe under the above conditions.

13.9 One hundred cu ft of air at 70°F dry bulb and 20 percent relative humidity are passed through a spray chamber where it comes in contact with water at the adiabatic-saturation temperature. It is cooled adiabatically and leaves at 53°F. What is the final percentage saturation of the air and how much moisture does it pick up?

13.10 In a drying process, air at 70°F and 60 percent relative humidity enters the dryer and is discharged at a temperature of 90°F and 90 percent relative humidity. How many cubic feet of air must be circulated through the dryer to remove 10 lb of water per min from the material to be dried? Note that a cubic foot of entering air has a greater volume on leaving.

13.11 An air-conditioning engineer has been given the following data on a room to be air conditioned:

Room sensible heat gain 40,000 Btu per hr

Room latent heat gain   8000 Btu per hr

Outside conditions   95°F dry bulb and 75°F wet bulb

Inside conditions   80°F dry bulb and 50 percent relative humidity

a. Calculate the refrigeration tonnage required with 10 percent safety factor.
b. Calculate the dry- and wet-bulb temperature of a mixture of 25 percent outside air and 75 percent return air.
c. Calculate the room sensible heat factor and apparatus dew point.
d. Calculate the cfm supply of air to satisfy room requirements.

13.12 A conditioner cooling surface fin section has a total capacity of 8.99 TR with an SHF of 0.67. With an entering air condition of 80°F dry-bulb temperature and 50 percent relative humidity and air flow rate of 200 lb per min, find the leaving air conditions of (a) dry-bulb temperature, (b) dew point, (c) wet-bulb temperature, and (d) relative humidity.

13.13 It is desired to condition 10,000 cfm of air to a leaving dew-point temperature corresponding to 20 grains per lb dry air by means of silica gel adsorption beds. Air enters the beds at 65°F dry-bulb temperature and 55° dew-point temperature. The heat of adsorption is plus 200 Btu per lb of vapor adsorbed. The following information is desired:
a. What is the outlet dry-bulb temperature?
b. If reactivation reduces the moisture content of the silica gel to 5 percent by weight, what is the weight of silica gel that is required in the beds

to permit extracting the required moisture on a 4-hr cycle and also limits the total moisture gain in the beds to 20 percent by weight for each cycle?

c. The beds are heated to 150°F by reactivation and are cooled back to 80°F by circulating dry air through them; air enters at 60°F and leaves the beds at 70°F. How much cooling air is necessary if the cooling cycle is to be restricted to 1 hr? The specific heat of silica gel may be taken as 0.22 Btu/(lb) (°F).

13.14 Outside air at 88°F dbt and 75 percent RH and at 29.92 in. Hg barometric pressure is passed through an air washer supplied with cooled water. It is then reheated to 68°F dbt and it is desired that the RH should be 40 percent. Determine the temperature of the air leaving the washer. Also determine, per 1000 cf of outside air, the pounds of water and the Btu of heat removed in the washer and the Btu added in the reheater.

13.15 A cooling tower installation is to be made near the top of a high mountain where the average atmospheric pressure is 13 psia. Four cfs of water is to be cooled from 105 to 82°F. Air enters the tower at 96°F at a partial vapor pressure of 0.75 psia and leaves as saturated air at 102°F. Calculate (a) the weight of air required per minute, and (b) the make-up water in pounds per hour. Note that the standard psychrometric chart cannot be used in the solution of this problem.

13.16 How many cubic feet of air per minute can be cooled with a 10-ton Freon-12 air-conditioning system from an initial temperature of 90°F dbt and 75 percent relative humidity to a final condition of 75°F dbt and 45 percent relative humidity? High-side temperature is 110°F and low-side is 40°F.

13.17 A high-pressure air-conditioning system is undergoing analysis. Fan ft³/min = 28,000; static pressure = 5.5 in. WG; fan outlet velocity = 2800 ft/min. System is a draw-through configuration.
a. What is the temperature rise across the fan?
b. How many tons of refrigeration are needed to offset this temperature rise?

13.18 The outside design temperature is 100°F and relative humidity is 15 percent. It is required to cool 1500 ft³/min of the air to the lowest possible temperature by evaporative cooling with the relative humidity of the air increasing to 60 percent. Determine how many pounds of water per minute are required and the final temperature.

SOLUTIONS

13.1 From the *Guide* or any other source, the conductivities of the various resistances are found and set up.

U_g (single glass)	1.13 Btu/(hr)(sq ft) (°F)
k for face brick	9.2 Btu/(hr)(sq ft) (°F) per in. thickness
k for common brick	5.0
k for fiberboard	0.34

k for plaster	8.00
k for gypsum lath	3.30
Coefficient air film inside	1.1
Coefficient air film outside (15-mph wind)	6.00
Coefficient air film inside room	1.65

Because U = summation of reciprocal of resistances in series, calculate each and then add.

$$U = \frac{1}{6.2015} = 0.161$$

Outside air film = 1/600	= 0.167
Face brick = 4/9.2	= 0.434
Common brick = 4/5	= 0.800
Air space = 1/1.1	= 0.909
Fiberboard = 0.75/0.34	= 2.200
Air space = 1/1.1	= 0.909
Gypsum lath = 0.375/3.3	= 0.114
Cement plaster = 0.5/8.00	= 0.0625
Inside air film = 1/1.65	= 0.606
Summation =	6.2015

$$\text{Wall area (gross)} = 20 \times 10 = 200 \text{ sq ft}$$

$$\text{Glass area} = 20 \text{ sq ft}$$

$$\text{Net wall area} = 180 \text{ sq ft}$$

Heat required to raise infiltration air = $10 \times 1.08 \times (70 - 0) = 756$ Btu per hr

a. Room heat loss:

Loss through wall = $0.161 \times 180 \times 70 = 2100$ Btu per hr
Loss through glass = $1.13 \times 20 \times 70 = 1580$
Infiltration loss = 756
Total loss = 4436 Btu per hr

b. From steam tables heat of condensation (H_{fg}) of steam at 2 psig is 966.1 Btu per lb. The steam rate is

$$\text{lb steam per hr} = \frac{\text{total loss}}{966.1} = \frac{4436}{966.1} = 4.64, \text{ say 5 lb per hr.}$$

c.

$$\text{cfm air at } 100°\text{F} = \frac{4436}{1.08(100 - 70)} = 137 \text{ cfm}$$

d. Because summation of resistances is previously found to be 6.2015 and the sum of the resistances measured from the inside to the air space is found to be 0.7825 (0.114 + 0.0625 + 0.606), the temperature within the $3\frac{5}{8}$-in. air space is taken from the *Guide* to be

$$t_i - (t_i - t_0)\frac{0.7825}{6.2015} = 70 - (70 - 0)\frac{0.7825}{6.2015} = 70 - 8.18 = 61.8°F.$$

13.2 See Exhibit 13.2. Overall coefficient for roof is given as 0.61. Because attic temperature is unknown and will be useful in determining dew point, let this be denoted by t_x. When equilibrium sets, i.e., when heat gain through ceiling into attic *equals* heat loss from attic through the roof to the outside, the temperature t_x may be determined. Heat gain through ceiling into attic is

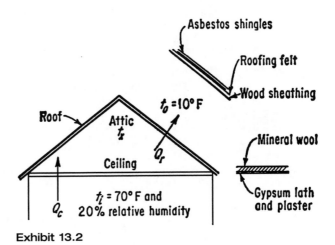

Exhibit 13.2

$$Q_c = U_c A_c(t_i - t_x).$$

Loss through the roof to outside is

$$Q_r = U_r A_r(t_x - t_0).$$

$$U_c = \frac{1}{\frac{1}{1.65} + \frac{1}{1.80} + \frac{2}{0.27} + \frac{1}{1.65}} = 0.109 \text{ Btu/(hr)(sq ft)(°F)}$$

$$Q_c = 0.109 \times 600(70 - t_x)$$
$$Q_r = 0.61 \times 1000(t_x - 10)$$
$$Q_c = Q_r = 0.109 \times 600(70 - t_x) = 0.61 \times 1000(t_x - 10)$$

from which t_x is calculated to be equal to 15.9°F, say 16°F. If this temperature is below the dew point corresponding to the room conditions, then condensation will occur on the attic side of the sheathing, provided there is no vapor seal on the attic side of the mineral wool insulation. If the temperature is above the dew point, then condensation will not occur. From the psychrometric chart, dew point corresponding to a condition of 70°F dry bulb and 20 percent relative humidity is 20°F. Without vapor seals, condensation will occur.

13.3   Pressures are so low that the perfect-gas law may be considered valid. With the use of steam tables we see that at 32°F vapor pressure of moisture = 0.180 in. Hg. The partial pressure of moisture in the air at 60 percent relative humidity is $0.60 \times 0.180 = 0.108$ in. Hg. The humidity ratio becomes

$$\frac{0.108}{29.92 - 0.108} \times 0.622 = 0.00226 \text{ lb water per lb dry air.}$$

This is also equivalent to

$$\frac{0.00226}{1 + 0.00226} = 0.00224 \text{ lb water per lb wet mixture.}$$

At 70°F vapor pressure of water is 0.739 in. Hg. The partial pressure of moisture in air at 30 percent relative humidity is

$$0.30 \times 0.739 = 0.221 \text{ in. Hg.}$$

The new humidity ratio becomes

$$\frac{0.221}{29.92 - 0.221} \times 0.662 = 0.00463 \text{ lb water per lb dry air.}$$

This is now also equivalent to

$$\frac{0.00463}{1 + 0.00463} = 0.00462 \text{ lb water per lb wet mixture.}$$

Water to be added is, therefore,

$$500,000 \times 3 \times (0.00462 - 0.00224)/13.8 = 258 \text{ lb per hr}$$

or $\qquad\qquad\qquad\qquad\qquad\qquad\qquad\qquad 258/500 = 0.515 \text{ gpm.}$

13.4   The total heat loss for the season is given by the formula

$$H' = \frac{H(t_r - t_m)N}{t_r - t_0} = \frac{24DH}{t_r - t_0}$$

where

H' = total loss, Btu for season
H = calculated heat loss, Btu per hr for estimated temperature difference $(t_r - t_0)$
N = number of hours that heating is required and for which t_m is average temperature
D = number of degree-days in the heating season
t_r = desired room temperature, °F
t_0 = outside temperature assumed for heat calculation H, °F
t_m = mean outside temperature for season of N hr, °F

The question implies the use of the second portion of the preceding equation.

$$H' = \frac{24 \times 7000 \times 240,000}{70 - 0} = 576 \times 10^6 \text{ Btu}$$

The amount of fuel used for the season is given by

$$F = \frac{H'}{eh} = \frac{576 \times 10^6}{0.60 \times 12,000} = 80,000 \text{ lb coal.}$$

$$\text{Cost} = 80,000/2000 \times 15 = \$600$$

13.5   a. Room air constitutes 99 parts having a specific gravity of 1.0, and vapor constitutes 1 part having a specific gravity of 5.0. Then

$$\begin{array}{r}0.99 \times 1.0 = 0.99 \\ 0.01 \times 5.0 = \underline{0.05} \\ 1.04\end{array}$$

This is the effective specific gravity of the room air mixture of air and vapor. Therefore, the room air mixture, compared with incoming outside air, would tend to move downward to the floor ever so slightly, expressed by the ratio of 104:100 and not by the ratio of 5:1, as is so frequently implied. This means that, in industry, the effects of drafts, window leakage and ventilation, plant traffic disturbances, or convection currents set up by process heat can easily dwarf into insignificance the effect of pure vapor specific gravity.

     b. Exhaust ft^3/min = (20 × 40 × 15)(15/60) = 3000 ft^3/min

     c. Locate exhaust points at floor level to avoid drawing room air across the workers' breathing zone.

     d. Because this is general (dilution) ventilation, room air-vapor concentration is the same as within the duct air stream.

     e. $\text{Bhp} = \dfrac{\text{air hp}}{\text{fan efficiency}} = \dfrac{0.16 \times 1.5 \times 3000}{1000 \times 0.6} = 1.2 \text{ bhp}$ Use a 2-hp motor.

     f. Makeup air heating load is obtained from

$$3000 \times 1.08 \times (80 - 10) = 226,800 \text{ Btu/h}$$

This is equivalent to the consumption of 230 lb steam per hour or 230 ft^3/h of a gaseous fuel having a gross heating value of 1000 Btu/ft^3.

13.6   This is a gas-dilution problem, the solution being straightforward if you remember that gas analyses are based on percent by volume; i.e., for perfect-gas mixtures, volume percent is proportional to the mols of each gas. Molecular weight of ethyl ether is 74 (C_2H_5—O—C_2H_5). Then the mols of ether removed per hour is 46/74 = 0.622.

    Because the LEL is given as 1.85 percent ether by volume, this is the same as saying 1.85 mol percent ethyl ether in the exhaust air leaving the dryer.

For each 1.85 mols ether exhausted, 98.15 mols air must dilute it. Then, the leaving air with 0.622 mol ethyl ether is

$$\frac{0.622}{0.0185} = x \quad \text{and} \quad x = 33 \text{ mols air/h}$$

At standard conditions of 60°F and 14.7 lb/in.^{2} abs the molal volume is 379 ft^3/mol. Then at a room temperature of 75°F the volume per mol of air is

$$\frac{379(460+75)}{460+60} = 390 \text{ ft}^3/\text{mol.}$$

Finally, the volume of air being exhausted is found to be

$$(33)(390) = 12,870 \text{ ft}^3/\text{h}$$

And this is the same as the volume being supplied. This amount of air is the least permissible to prevent the explosive limit from being attained in the main bulk of the exhaust system. However, local concentrations within the dryer itself may be explosive, so that a sparkproof design is a requirement.

13.7   Enter chart in Exhibit 13.7 at 98.8 mm Hg. At the 40°C curve turn down and read 8.8 SO$_2$ per 1000 g water and

$$(8.8/1000)(2.205/2.205) = 19.4 \text{ lb SO}_2/2205 \text{ lb H}_2\text{O}$$
$$= 0.0088 \text{ lb/lb} \quad \text{or} \quad 0.88 \text{ lb SO}_2/100 \text{ lb H}_2\text{O.}$$

In 100 ft^3 "rich" gas, there are

$$(100/459)(0.87)(273/313)(29) = 4.79 \text{ lb air}$$
$$(100/359)(0.13)(273/313)(64) = 2.02 \text{ lb SO}_2.$$

On an hourly basis,

$$(4.79)(60) = 287.4 \text{ lb air/h} = G$$
$$(2.02)(60) = 121.2 \text{ lb SO}_2/\text{h.}$$

Partial pressure SO$_2$ in the entering gas stream is $(0.13)(760) = 98.8$ mm Hg.

Unabsorbed SO$_2$ in the waste gas leaving the absorber is $(0.01)(121.2) = 1.21$ lb SO$_2$ per hour.

From Exhibit 13.7, it has been found that the concentration of an aqueous solution of SO$_2$ at a partial pressure of 98.8 mm Hg at 40°C is 0.88 lb SO$_2$ per 100 lb water. This is the maximum strength of solution that can be produced from the entering gas. Then,

$$L(X_1 - X_0) = G(Y_1 - Y_0) = L(0.0088 - 0)$$
$$= 287.4(121.2/287.4 - 1.2/287.4)$$
$$0.0088 \, L = 121.2 - 1.2 = 120 \quad \text{and} \quad L$$
$$= 13,600 \text{ lb water/h}$$

13.8   a.   Set up the following tabulation:

ah	$V = 4005\sqrt{h}$, ft/min	Ah	$V = 4005\sqrt{h}$, ft/min
0.10	1266	0.14	1498
0.11	1328	0.18	1699
0.19	1746	0.22	1879
0.27	2081	0.30	2193
0.35	2369	0.35	2369
0.34	2335	0.36	2403
0.29	2157	0.24	1962
0.22	1879	0.19	1746
0.16	1602	0.14	1498
0.13	1444	0.11	1328
Totals	18,207		18,575

$$\text{Average} = (18,207 + 18,575)/20 \text{ readings}$$
$$= 1839 \text{ ft/min ft}^3/\text{min flow} = VA$$
$$= 1839 \times 0.7854(12/12)^2 = 1444 \text{ ft}^3/\text{min}$$

b.   Finding C_e, knowing the velocity and the throat suction at inlet:

$$V = 4005C_e\sqrt{h}$$

from which $C_e = (V)/\left(4005\sqrt{h}\right) = (1839)/\left(4005 \times \sqrt{0.44}\right) = 0.69$

c.   New airflow:

$$\text{ft}^3/\text{min} = 4005\ C_e\ \sqrt{h}A$$
$$= 4005(0.69)(0.53)(0.7854) = 1150 \text{ ft}^3/\text{min}$$

13.9   See Exhibit 13.9 skeleton psychrometric chart. Point A gives the value of 0.00316 lb of water per lb of dry air, or

$$0.00316 \times 7,000 = 22.12 \text{ grains.}$$

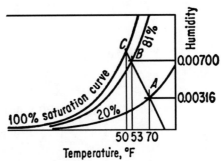

Exhibit 13.9

By following along the adiabatic-saturation line to a dry-bulb temperature of 53°F (point *B*), a final humidity of 0.0070 and a percentage saturation (relative humidity) of 81 are obtained. The adiabatic-saturation temperature is about 50°F.

From the psychrometric chart by interpolation the volume of air at 70°F and 20 percent relative humidity is found to be 13.45 cu ft per lb of dry air. Therefore, total weight of dry air is

$$100/13.45 = 7.44 \text{ lb.}$$

And the weight of water picked up by the air is

$$7.44(0.00700 - 0.00316) = 0.0286 \text{ lb.}$$

13.10
$$0.60 = \frac{\text{wt vapor per cu ft actual}}{\text{wt vapor per cu ft in air saturated with vapor}}$$

$$0.60 = \frac{x}{\frac{1}{869}} \quad x = \frac{0.60}{869} = 0.691 \times 10^{-3} \text{ lb at entrance}$$

For leaving condition if volume is the same as at entrance condition,

$$0.90 = \frac{y}{1/468.5} \quad y = \frac{0.90}{468.5} = 1.92 \times 10^{-3} \text{ lb.}$$

For every cubic foot entering there is

$$1 \times \frac{460 + 90}{460 + 70} = 1.039 \text{ cu ft leaving.}$$

That is to say, for every cubic foot entering 1.039 cu ft leaves due to increase in temperature. Then the weight of vapor leaving is found by

$$0.00192 \times 1.039 = 0.00199468 \text{ lb}$$

Thus, there is removed for each cubic foot of air passing through the dryer: $0.001995 - 0.000691 = 0.001304$ lb vapor. Finally, by dimensional analysis,

$$\frac{\text{lb water}}{\text{per min}} \times \frac{1}{\text{lb per cu ft removed}} = \frac{10}{0.001304} = 7660 \text{ cfm.}$$

13.11 a.
$$\frac{(40,000 + 8000)1.10}{12,000} = 4.4 \text{ TR}$$

b. Dry-bulb temperature:

$$\text{Outside air} = 0.25 \times 95 = 23.8°F$$
$$\text{Return air} = 0.75 \times 80 = \underline{60.0°F}$$
$$\text{Mixture temperature} = 83.8°F$$

Wet-bulb temperature: On psychrometric chart draw a straight line connecting the outside air condition and the inside air condition (see Exhibit 13.11). Locate the calculated dry-bulb temperature of 83.8°F along the base of the chart and move up in a straight line until you hit the straight line just drawn. Then read up and to the left along constant wet-bulb line the wet-bulb temperature of 69°F.

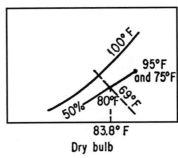

Exhibit 13.11

c. Room sensible heat factor (SHF):

$$SHF = \frac{\text{room sensible heat}}{\text{room sensible heat} + \text{room latent heat}}$$
$$= \frac{40,000}{40,000 + 8000} = 0.833$$

From the Carrier System Design Manual, Part 1, Load Estimating, pp. 1–147, the apparatus dew point may be calculated from the following equation:

$$SHF = \frac{0.244(t_{rm} - t_{adp})}{0.244(t_{rm} - t_{adp}) + 1076/7000(W_{rm} - W_{adp})}$$

where

W_{rm} = room moisture content, grains per lb dry air

W_{adp} = moisture content at apparatus dew point, grains per lb dry air

t_{rm} = room dry-bulb temperature, °F

t_{adp} = apparatus dew point, °F

0.244 = specific heat of moist air at 55°F dew point, Btu/(°F)(lb) of dry air

1076 = average heat removal required to condense 1 lb water vapor from room air

$0.244 (t_{rm} - t_{adp})$ = total sensible load

$\dfrac{1076}{7000}(W_{rm} - W_{adp})$ = total latent load

Of course, the above-mentioned Carrier Manual will give you the apparatus dew point (ADP) at once. But it is required that you calculate it. To do this, assume a value of ADP from the psychrometric chart, using the slope for the SHF directly. This assumption is found to be between 56 and 58°F. Try 57°F. Also from the psychrometric chart, W_{rm} and W_{adp} are found. Now insert these in the SHF equation, in its simplified form below.

$$SHF = \frac{1}{1+0.628\frac{W_{rm}-W_{adp}}{t_{rm}-t_{adp}}}$$

$$= \frac{1}{1+0.628\frac{77-69.5}{80-57}} = 0.831$$

Thus, the ADP assumed is close enough. Use 57°F.

d. Air supply, cfm. If we assume all air passing through the onditioner will come in contact with the cooling surface (Carrier zero bypass factor), then the cfm is given by

$$cfm = \frac{RSH}{1.08(t_{rm}-t_{adp})} = \frac{40,000}{1.08(80-57)} = 1610\ cfm$$

13.12   Set up the table of *entering conditions* as follows:

Dry-bulb temperature	80°F
Wet-bulb temperature	67°F
Dew-point temperature	60°F
Relative humidity	50 percent
Total heat (enthalpy)	31.15 Btu per lb
Grains moisture per lb dry air	77.3
Sp, vol, cu ft per lb	13.84
Air flow rate, lb per min	200
Sensible heat factor	0.67

a. $$0.67 = \frac{\text{total sensible heat}}{\text{sensible heat}+\text{latent}}$$

Total sensible heat = 0.67 × 8.99 = 6.03 TR. The temperature drop through the cooling surface is

$$\text{Temp. difference} = \frac{6.03\times200}{220\times0.24} = 22.8°F.$$

Therefore, the leaving dry-bulb temperature is 80 − 22.8 = 57.2°F.

b.   Total load minus sensible load equals moisture (latent) load, i.e.,

$$(8.999 \times 200) - (6.03 \times 200) = 592 \text{ Btu per min}$$

$$\frac{592 \times 7000}{1076 \times 220} = 17.4 \text{ grains per lb dry air removed by coil}$$

At 80°F dry-bulb temperature and 50 percent relative humidity the moisture content is 77.3 grains per lb dry air. Then the moisture content for the leaving condition is 77.3 − 17.4 = 59.9 grains per lb dry air. This corresponds approximately to a dew point of 53°F.

c.   Wet-bulb temperature. From psychrometric chart with use of leaving conditions of dry-bulb and wet-bulb temperatures read 54.9°F wet-bulb temperature.

d.   Relative humidity. Also from the chart, read 85 percent relative humidity.

13.13   This is considered chemical drying and is represented by a straight line along the wet-bulb temperature between the limits of the drying process *AB* only in case the drying is purely by adsorption (the drying agent does not dissolve in the water extracted from the air) and only in case the drying agent does not retain an appreciable amount of the heat of vaporization liberated when the water is condensed on the surface of the adsorber. In case an appreciable amount of heat of vaporization is retained by the adsorber, the process takes place along a line below the wet-bulb temperature *AB′*. If the drying agent is soluble in water (such as calcium chloride), the drying process takes place along a line which lies above *AB″* or below *AB′*, depending on whether heat is liberated or absorbed when the agent is dissolved in water (see Exhibit 13.13).

a.   From psychrometric chart at inlet air conditions of 65° dry-bulb temperature and 55° dew-point temperature the specific volume is 13.4 cu ft per lb and the moisture content is found to be 64 grains per lb. The change in temperature will depend on the moisture pickup converted to Btu. Knowing the cfm being handled and the Btu pickup the temperature rise can be calculated. The moisture pickup is 64 − 20 = 44 grains per lb of dry air. This converted to pounds per hour is given as

$$\frac{10,000 \times 60 \text{ min}}{13.4} = 44,800 \text{ lb per hr of dry air.}$$

The temperature rise is

$$\frac{44,800 \times 44 \times 200}{7000} \times \frac{1}{1.08} \times \frac{1}{10,000} = 5.22°F.$$

The leaving air temperature is $65 + 5.22 = 70.22°F$. Examination of Exhibit 13.13 and the psychrometric chart indicates that an appreciable amount of heat of vaporization has been retained by the silica gel.

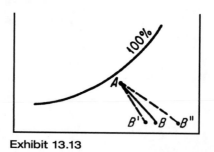

Exhibit 13.13

b. For the 4-hr cycle operation the total moisture removed is simply

$$\frac{44,800 \times 44 \times 4}{7000} = 1125 \text{ lb in 4 hr}$$

Problem says that the moisture gained percentage-wise is

$$20 - 5 = 15 \text{ percent.}$$

Thus, the weight of silica gel required is $1125/0.15 = 7500$ lb.

c. The heat picked up by the reactivation air is represented by the temperature drop.

$$7500 \times 0.22 \times (150 - 80) = 115,000 \text{ Btu per hr}$$

And the air required to accomplish this is $Q/[1.08(t_2 - t_1)]$.

$$\text{cfm} = 115,000/(1.08 \times 10) = 10,650 \text{ cfm}$$

13.14   Refer to Exhibit 13.14. Now list all data from psychrometric chart at points 1, 2, and 3. At point 2 air is assumed saturated, so its RH is 100 percent. At point 3, dpt is 43, which is also true at point 2, because no moisture is added in reheater (no dpt change). Also, at point 2, RH is 100 percent, so dbt and wbt are also $43°F$. Therefore, temperature of air leaving washer is $43°F$.

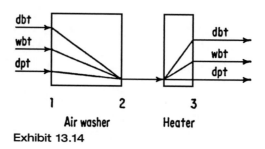

Exhibit 13.14

	1	2	3
dbt, °F	88	43	68
wbt, °F	81.2	43	54.3
dpt, °F	79.1	43	43
RH, %	75	100	40
S, gr per lb	151	41	41 dry air
H, Btu per lb	44.8	16.7	22.9 dry air
$\bar{V}$, cf per lb	14.28	12.78	13.42

(151 − 41)/7000 = 0.0157 lb vapor per lb dry air removed in washer
(0.0157/14.28) × 1000 = 1.1 lb moisture removed per 1000 cf outside air

44.8 − 16.7 = 28.1 Btu per lb dry air removed in washer

(28.1/14.28) × 1000 = 1960 Btu removed per 1000 cf outside air
22.9 − 16.7 = 6.2 Btu per lb dry air added in reheater

(6.2/14.28) × 1000 = 433 Btu added per 1000 cf outside air

13.15 Heat transferred from the water to the air:

$$Q = wC_p(T_1 - T_2) = 4 \times 62.5 \times 1 \times (105 - 82) = 5750 \text{ Btu per sec}$$

Heat absorbed by the air in the change in enthalpy of the air mixture from inlet to outlet:

$$Q = H_0 - H_i = w_a(h_0 - h_i) \qquad \text{where } h_i = \text{enthalpy of air "in"}$$

$$p_r = p_v + p_a = 13.0 \text{ psia}$$
$$p_a = 13 - 0.75 = 12.25 \text{ psia}$$

If we assume the vapor and air act as perfect gases, the humidity ratio is

$$w_i = \frac{p_v R_a}{p_a R_v} = \frac{0.75 \times 53.3}{12.25 \times 85.7} = 0.0381 \text{ lb vapor per lb dry air.}$$

Total weight of inlet air mixture w_r = 1.0000 + 0.0381 = 1.0381 lb
From Keenan and Keyes' Steam Tables, for an inlet temperature t_i = 96°F, h_g = 1103.5 Btu/lb. Now the enthalpy of the vapor is given by

$$h_g = w_v h_g = 0.0381 \times 1103.5 = 42.05.$$
Enthalpy of dry air is $h_a = C_p t_i = 0.24 \times 96 = 23.05.$
$$h_g + h_a = 42.05 + 23.05 = 65.10.$$

Enthalpy of air in:

$$h_i = (h_v + h_a)/1.0381 = 65.10/1.0381 = 63.0 \text{ Btu per lb}$$

Enthalpy of air out h_0. Because the outlet air is saturated at 102°F,

$$p_v = 1.0078 \text{ psia.}$$
$$p_a = p_r - p_v = 13 - 1.0078 = 11.9922 \text{ psia}$$

Humidity ratio:

$$w_0 = \frac{p_v R_a}{p_a R_v} = \frac{1.0078 \times 53.3}{11.9922 \times 85.7} = 0.0522 \text{ lb vapor per lb dry air}$$

From Keenan and Keyes' Steam Tables, for an outlet temperature $t_0 = 102°F$

$$h_g = 1106.1 \text{ Btu per lb}$$

Enthalpy of vapor:

$$h_v = w_v h_g = 0.0522 \times 1106.1 = 57.5$$

Enthalpy of dry air:

$$h_a = C_p t_0 = 0.24 \times 102 = 24.5$$

Then $h_v + h_a = 57.7 + 24.5 = 82.2$. The enthalpy of air out is

$$h_0 = (h_v + h_a)/1.0522 = 82.2/1.0522 = 78.0 \text{ Btu per lb.}$$

a. Weight of air (w_a) required per minute is

$$w_a = \frac{Q \times 60}{h_0 - h_i} = \frac{5750 \times 60}{78 - 63} = 23,000 \text{ lb per min}$$

b. Pounds per hour of makeup water. Each pound of air picks up an amount of water vapor equal to

$$\Delta w = w_a - w_i = 0.0522 - 0.0381 = 0.0141 \text{ lb vapor per lb air}$$
$$w_{H_2O} = 23,000 \times 60 \times 0.0141 = 19,500 \text{ lb/hr} \quad \text{or} \quad 19,500/500 = 39 \text{ gpm}$$

13.16 Cooling load is $10 \times 200 = 2000$ Btum. From the psychrometric chart, enthalpy per pound of air at entering conditions is 46.9 Btu; at leaving conditions it is 27.1 Btu per lb. Thus, heat removed per lb air = 46.9 − 27.1 = 19.8 Btu per lb. Weight of air flow per min = 2000/19.8 = 101 lb.

From the psychrometric chart, specific volume of air = 14.36 cu ft per lb. Finally, cfm = 101 × 14.36 = 1450. Note that specific volume is based on outside air conditions.

13.17 a. Temperature rise across fan is given by

$$\frac{5.19 \times (TP_2 - TP_1)}{778 \times C_p \times d_1 \times FTE}$$

where

TP_2 = downstream total pressure, in. WG

TP_1 = upstream total pressure, in. WG

FTE = fan total efficiency

C_p = specific heat of air = 0.24

d_1 = upstream air density, lb/ft^3.

Now continuing, fan outlet velocity = 2800 ft/min and the fan velocity pressure = $(2800/4005)^2$ = 0.49 in. WG. Now assume casing velocity pressure is negligible. Normally, it would be about 500 ft/min, which is equal to a velocity pressure of $(500/4005)^2$ = 0.0156 in. WG.

Downstream total pressure is 0.49 + 5.5 = 5.99 in. WG, say 6.00 in. WG = TP_2. Also assume a fan efficiency of 0.84, or 84 percent.

The total pressure on the upstream side of the fan is 0 in. WG, i.e., TP_1 = 0. *Note:* On any system where you have louvers, filters, and coils, TP_1 would be a negative number, as the pressure when related to atmospheric is negative. However, we have already calculated the required fan static pressure. Therefore, fan static plus velocity pressure at the outlet gives us $TP_2 - TP_1$, neglecting the velocity pressure in the inlet plenum. Let d_1 = 0.075 lb/ft^3 air density.

Finally, temperature rise across fan is

$$\frac{5.19 \times 6}{778 \times 0.24 \times 0.075 \times 0.84} = 2.65°F.$$

b. Tonnage required over and above system heat load

$$\frac{28,000 \times 1.08 \times 2.6}{12,000} = 6.7 \text{ tons}$$

13.18 Use the psychrometric chart for normal atmospheric conditions. Refer to Exhibit 13.18 and note point 1:

$$T_{dbl} = 100°F \quad RH_1 = 15 \text{ percent}$$
$$T_{wb} = 66°F \quad SH_1 = 42 \text{ gr/lb dry air}$$

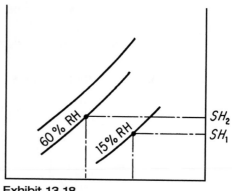

Exhibit 13.18

Assume water supply is at 66°F or equal to T_{wb}. Then follow the constant wet-bulb line at 66°F from $RH_1 = 15$ percent to $RH_2 = 60$ percent and read $T_{db2} = 75.7$°F final temperature.

Also read $SH_2 = 80$ gr/lb dry air and $V_2 = 13.74$ ft^3/lb dry air.

Mass of airflow in = 1500/14.24 = 105.2 lb/min

Amount of water in incoming air = 105.2 × 42 = 4418 gr/min

Amount of water in air at final condition = 105.2 × 80 = 8416 gr/min

Amount of water added to airstream = 8416 − 4418 = 3998 gr/min

Finally, rate of water added = 3998/7000 = 0.571 lb min

Engineering Economics

PROBLEMS

A.1 A loan was made 2½ years ago at 8% simple annual interest. The principal amount of the loan has just been repaid along with $600 of interest. The principal amount of the loan was closest to
a. $300   c. $4000
b. $3000   d. $5000

A.2 A $1000 loan was made at 10% simple annual interest. It will take how many years for the amount of the loan and interest to equal $1700?
a. 6 years   c. 8 years
b. 7 years   d. 9 years

A.3 A retirement fund earns 8% interest, compounded quarterly. If $400 is deposited every three months for 25 years, the amount in the fund at the end of 25 years is nearest to
a. $50,000   c. $100,000
b. $75,000   d. $125,000

A.4 For some interest rate i, and some number of interest periods n, the uniform series capital recovery factor is 0.2091 and the sinking fund factor is 0.1941. The interest rate i must be closest to
a. 1½%   c. 3%
b. 2%   d. 4%

A.5 The repair costs for some handheld equipment is estimated to be $120 the first year, increasing by $30 per year in subsequent years. The amount a person will need to deposit into a bank account paying 4% interest to provide for the repair costs for the next five years is nearest to
a. $500   c. $700
b. $600   d. $800

A.6 An "annuity" is defined as the
  a. earned interest due at the end of each interest period
  b. cost of producing a product or rendering a service
  c. total annual overhead assigned to a unit of production
  d. series of equal payments occurring at equal periods of time

A.7 One thousand dollars is borrowed for one year at an interest rate of 1% per month. If this same sum of money is borrowed for the same period at an interest rate of 12% per year, the saving in interest charges is closest to
  a. $0   c. $7
  b. $5   d. $14

A.8 How much should a person invest in a fund that will pay 9%, compounded continuously, if he wishes to have $10,000 in the fund at the end of 10 years? The amount is nearest to
  a. $4000   c. $6000
  b. $5000   d. $7000

A.9 A store charges $1\frac{1}{2}\%$ interest per month on credit purchases. This is equivalent to a nominal annual interest rate of
  a. 1.5%   c. 18.0%
  b. 15.0%   d. 19.6%

A.10 A small company borrowed $10,000 to expand its business. The entire principal of $10,000 will be repaid in two years, but quarterly interest of $330 must be paid every three months. The nominal annual interest rate the company is paying is closest to
  a. 3.3%   c. 6.6%
  b. 5.0%   d. 13.2%

A.11 A store policy is to charge 3% interest every two months on the unpaid balance in charge accounts. The effective interest rate is closest to
  a. 6%   c. 15%
  b. 12%   d. 19%

A.12 The effective interest rate is 19.56%. If there are 12 compounding periods per year, the nominal interest rate is closest to
  a. 1.5%   c. 9.0%
  b. 4.5%   d. 18.0%

A.13 A deposit of $300 was made one year ago into an account paying monthly interest. If the account now has $320.52, the effective annual interest rate is closest to
  a. 7%   c. 12%
  b. 10%   d. 15%

A.14 In a situation where the effective interest rate per year is 12%, based on monthly compounding, the nominal interest rate per year is closest to
  a. 8.5%   c. 10.0%
  b. 9.3%   d. 11.4%

A.15 If 10% nominal annual interest is compounded daily, the effective annual interest rate is nearest to
  a. 10.00%   c. 10.50%
  b. 10.38%   d. 10.75%

A.16 If 10% nominal annual interest is compounded continuously, the effective annual interest rate is nearest to
  a. 10.00%   c. 10.50%
  b. 10.38%   d. 10.75%

A.17 If the quarterly effective interest rate is $5\frac{1}{2}$% with continuous compounding, the nominal interest rate is nearest to
  a. 5.5%   c. 16.5%
  b. 11.0%   d. 21.4%

A.18 A continuously compounded loan has what effective interest rate if the nominal interest rate is 25%?
  a. $e^{1.25}$ c. $e^{0.25} - 1$
  b. $e^{0.25}$ d. $\ln(1.25)$

A.19 A continuously compounded loan has what *nominal interest rate* if the *effective interest rate* is 25%?
  a. $e^{1.25}$ c. $\ln(1.25)$
  b. $e^{0.25}$ d. $\log_{10}(1.25)$

A.20 An individual wishes to deposit a certain quantity of money now so that he will have $500 at the end of five years. With interest at 4% per year, compound semiannually, the amount of the deposit is nearest to
  a. $340   c. $410
  b. $400   d. $608

A.21 A steam boiler is purchased on the basis of guaranteed performance. A test indicates that the operating cost will be $300 more per year than the manufacturer guaranteed. If the expected life of the boiler is 20 years, and money is worth 8%, the amount the purchaser should deduct from the purchase price to compensate for the extra operating cost is nearest to
  a. $2950   c. $4100
  b. $3320   d. $5520

A.22 A consulting engineer bought a fax machine with one year's free maintenance. In the second year the maintenance is estimated at $20. In subsequent years the maintenance cost will increase $20 per year (that is, third year maintenance will be $40, fourth year maintenance will be $60, and so forth). The amount that must be set aside now at 6% interest to pay the maintenance costs on the fax machine for the first six years of ownership is nearest to
  a. $101   c. $229
  b. $164   d. $284

A.23 An investor is considering buying a 20-year corporate bond. The bond has a face value of $1000 and pays 6% interest per year in two semiannual payments. Thus the purchaser of the bond will receive $30 every six months, and in addition he will receive $1000 at the end of 20 years, along with the last $30 interest payment. If the investor believes he should receive 8% annual interest, compounded semiannually, the amount he is willing to pay for the bond value of closest to
a. $500   c. $800
b. $600   d. $900

A.24 Annual maintenance costs for a particular section of highway pavement are $2000. The placement of a new surface would reduce the annual maintenance cost to $500 per year for the first five years and to $1000 per year for the next five years. The annual maintenance after 10 years would again be $2000. If maintenance costs are the only saving, the maximum investment that can be justified for the new surface, with interest at 4%, is closest to
a. $5,500   c. $10,000
b. $7,170   d. $10,340

A.25 A project has an initial cost of $10,000, uniform annual benefits of $2400, and a salvage value of $3000 at the end of its 10-year useful life. At 12% interest the net present worth of the project is closest to
a. $2,500   c. $4,500
b. $3,500   d. $5,500

A.26 A person borrows $5000 at an interest rate of 18%, compounded monthly. Monthly payments of $167.10 are agreed upon. The length of the loan is closest to
a. 12 months   c. 24 months
b. 20 months   d. 40 months

A.27 A machine costing $2000 to buy and $300 per year to operate will save labor expenses of $650 per year for eight years. The machine will be purchased if its salvage value at the end of eight years is sufficiently large to make the investment economically attractive. If an interest rate of 10% is used, the minimum salvage value must be closest to
a. $100   c. $300
b. $200   d. $400

A.28 The amount of money deposited 50 years ago at 8% interest that would now provide a perpetual payment of $10,000 per year is nearest to
a. $3,000   c. $50,000
b. $8,000   d. $70,000

A.29 An industrial firm must pay a local jurisdiction the cost to expand its sewage treatment plant. In addition, the firm must pay $12,000 annually toward the plant operating costs. The industrial firm will pay sufficient money into a fund that earns 5% per year, to pay its share of the plant operating costs forever. The amount to be paid to the fund is nearest to
a. $15,000   c. $160,000
b. $60,000   d. $240,000

A.30 At an interest rate of 2% per month, money will double in value in how many months?

a. 20 months   c. 50 months
b. 35 months   d. 65 months

A.31 A woman deposited $10,000 into an account at her credit union. The money was left on deposit for 80 months. During the first 50 months the woman earned 12% interest, compounded monthly. The credit union then changed its interest policy so that the woman earned 8% interest compounded quarterly during the next 30 months. The amount of money in the account at the end of 80 months is nearest to

a. $10,000   c. $20,000
b. $15,000   d. $25,000

A.32 An engineer deposited $200 quarterly in her savings account for three years at 6% interest, compounded quarterly. Then for five years she made no deposits or withdrawals. The amount in the account after eight years is closest to

a. $1200   c. $2400
b. $1800   d. $3600

A.33 A sum of money, Q, will be received six years from now. At 6% annual interest the present worth now of Q is $60. At this same interest rate the value of Q 10 years from now is closest to

a. $60   c. $90
b. $77   d. $107

A.34 If $200 is deposited in a savings account at the beginning of each of 15 years and the account earns interest at 6%, compounded annually, the value of the account at the end of 15 years will be most nearly

a. $4500   c. $4900
b. $4700   d. $5100

A.35 The maintenance expense on a piece of machinery is estimated as follows:

Year	1	2	3	4
Maintenance	$150	$300	$450	$600

If interest is 8%, the equivalent uniform annual maintenance cost is closest to

a. $250   c. $350
b. $300   d. $400

A.36 A payment of $12,000 six years from now is equivalent, at 10% interest, to an annual payment for eight years starting at the end of this year. The annual payment is closest to

a. $1000   c. $1400
b. $1200   d. $1600

A.37 A manufacturer purchased $15,000 worth of equipment with a useful life of six years and a $2000 salvage value at the end of the six years. Assuming a 12% interest rate, the equivalent uniform annual cost is nearest to
  a. $1500   c. $3500
  b. $2500   d. $4500

A.38 Consider a machine as follows:

Initial cost: $80,000

End-of-useful-life salvage value: $20,000

Annual operating cost: $18,000

Useful life: 20 years

Based on 10% interest, the equivalent uniform annual cost for the machine is closest to
  a. $21,000   c. $25,000
  b. $23,000   d. $27,000

A.39 Consider a machine as follows:

Initial cost: $80,000

Annual operating cost: $18,000

Useful life: 20 years

What must be the salvage value of the machine at the end of 20 years for the machine to have an equivalent uniform annual cost of $27,000? Assume a 10% interest rate. The salvage value is closest to
  a. $10,000   c. $40,000
  b. $20,000   d. $50,000

A.40 Twenty-five thousand dollars is deposited in a savings account that pays 5% interest, compounded semiannually. Equal annual withdrawals are to be made from the account beginning one year from now and continuing forever. The maximum amount of the equal annual withdrawals is closest to
  a. $625   c. $1250
  b. $1000   d. $1265

A.41 An investor is considering investing $10,000 in a piece of land. The property taxes are $100 per year. The lowest selling price the investor must receive if she wishes to earn a 10% interest rate after keeping the land for 10 years is
  a. $21,000   c. $27,000
  b. $23,000   d. $31,000

A.42 The rate of return of a $10,000 investment that will yield $1000 per year for 20 years is closest to
  a. 1%   c. 8%
  b. 4%   d. 12%

A.43   An engineer invested $10,000 in a company. In return he received $600 per year for six years and his $10,000 investment back at the end of the six years. His rate of return on the investment was closest to
a. 6%   c. 12%
b. 10%   d. 15%

A.44   An engineer made 10 annual end-of-year purchases of $1000 of common stock. At the end of the tenth year, just after the last purchase, the engineer sold all the stock for $12,000. The rate of return received on the investment is closest to
a. 2%   c. 8%
b. 4%   d. 10%

A.45   A company is considering buying a new piece of machinery.

Initial cost: $80,000

End-of-useful-life salvage value: $20,000

Annual operating cost: $18,000

Useful life: 20 years

The machine will produce an annual savings in material of $25,700. What is the before-tax rate of return if the machine is installed? The rate of return is closest to
a. 6%   c. 10%
b. 8%   d. 15%

A.46   Consider the following situation: Invest $100 now and receive two payments of $102.15—one at the end of Year 3, and one at the end of Year 6. The rate of return is nearest to
a. 8%   c. 18%
b. 12%   d. 22%

A.47   Two mutually exclusive alternatives are being considered:

Year	A	B
0	−$2500	−$6000
1	+746	+1664
2	+746	+1664
3	+746	+1664
4	+746	+1664
5	+746	+1664

The rate of return on the difference between the alternatives is closest to
a. 6%   c. 10%
b. 8%   d. 12%

A.48   A project will cost $50,000. The benefits at the end of the first year are estimated to be $10,000, increasing $1000 per year in subsequent years. Assuming a 12% interest rate, no salvage value, and an eight-year analysis period, the benefit-cost ratio is closest to

a. 0.78   c. 1.28
b. 1.00   d. 1.45

A.49   Two alternatives are being considered:

	A	B
Initial cost	$500	$800
Uniform annual benefit	$140	$200
Useful life, years	8	8

The benefit-cost ratio of the difference between the alternatives, based on a 12% interest rate, is closest to

a. 0.60   c. 1.00
b. 0.80   d. 1.20

A.50   An engineer will invest in a mining project if the benefit-cost ratio is greater than one, based on an 18% interest rate. The project cost is $57,000. The net annual return is estimated at $14,000 for each of the next eight years. At the end of eight years the mining project will be worthless. The benefit-cost ratio is closest to

a. 1.00   c. 1.21
b. 1.05   d. 1.57

A.51   A city has retained your firm to do a benefit-cost analysis of the following project:

Project cost: $60,000,000

Gross income: $20,000,000 per year

Operating costs: $5,500,000 per year

Salvage value after 10 years: None

The project life is 10 years. Use 8% interest in the analysis. The computed benefit-cost ratio is closest to

a. 0.80   c. 1.50
b. 1.00   d. 1.60

A.52   A piece of property is purchased for $10,000 and yields a $1000 yearly profit. If the property is sold after five years, the minimum price to break even, with interest at 6%, is closest to

a. $5000   c. $7700
b. $6500   d. $8300

A.53   Given two machines:

	A	B
Initial cost	$55,000	$75,000
Total annual costs	$16,200	$12,450

With interest at 10% per year, at what service life do these two machines have the same equivalent uniform annual cost? The service life is closest to

a. 5 years   c. 7 years
b. 6 years   d. 8 years

A.54   A machine part that is operating in a corrosive atmosphere is made of low-carbon steel. It costs $350 installed, and lasts six years. If the part is treated for corrosion resistance it will cost $700 installed. How long must the treated part last to be as economic as the untreated part, if money is worth 6%?

a. 8 years   c. 15 years
b. 11 years   d. 17 years

A.55   A firm has determined the two best paints for its machinery are Tuff-Coat at $45 per gallon and Quick at $22 per gallon. The Quick paint is expected to prevent rust for five years. Both paints take $40 of labor per gallon to apply, and both cover the same area. If a 12% interest rate is used, how long must the Tuff-Coat paint prevent rust to justify its use?

a. 5 years   c. 7 years
b. 6 years   d. 8 years

A.56   Two alternatives are being considered:

	A	B
Cost	$1000	$2000
Useful life in years	10	10
End-of-useful-life salvage value	100	400

The net annual benefit of A is $150. If interest is 8%, what must be the net annual benefit of B for the two alternatives to be equally desirable? The net annual benefit of B must be closest to

a. $150   c. $225
b. $200   d. $275

A.57   Which one of the following is *NOT* a method of depreciating plant equipment for accounting and engineering economics purposes?
a. double-entry method
b. modified accelerated cost recovery system
c. sum-of-years-digits method
d. straight-line method

A.58   A machine costs $80,800, has a 20-year useful life, and an estimated $20,000 end-of-useful-life salvage value. Assuming sum-of-years-digits depreciation, the book value of the machine after two years is closest to
a. $21,000   c. $59,000
b. $42,000   d. $69,000

A.59   A machine costs $100,000. After its 25-year useful life, its estimated salvage value is $5,000. Based on double-declining-balance depreciation, what will be the book value of the machine at the end of three years? The book value is closest to
a. $16,000   c. $58,000
b. $22,000   d. $78,000

QUESTIONS A.60 TO A.63

Special tools for the manufacture of finished plastic products cost $15,000 and have an estimated $1000 salvage value at the end of an estimated three-year useful life.

A.60   The third-year straight line depreciation is closest to
a. $3000   c. $4000
b. $3500   d. $4500

A.61   The first year modified-accelerated-cost-recovery-system (MACRS) depreciation is closest to
a. $3000   c. $5000
b. $4000   d. $6000

A.62   The second-year sum-of-years-digits (SOYD) depreciation is closest to
a. $3000   c. $4000
b. $3500   d. $4500

A.63   The second-year sinking-fund depreciation, based on 8% interest, is nearest to
a. $3000   c. $4000
b. $3500   d. $4500

A.64   An individual who has a 28% incremental income tax rate is considering purchasing a $1000 taxable corporation bond. As the bondholder he will receive $100 a year in interest and his $1000 back when the bond becomes due in six years. This individual's after-tax rate of return from the bond is nearest to
a. 6%   c. 8%
b. 7%   d. 9%

A.65   A $20,000 investment in equipment will produce $6000 of net annual benefits for the next eight years. The equipment will be depreciated by straight-line depreciation over its eight-year useful life. The equipment has no salvage value. Assuming a 34% income tax rate, the after-tax rate of return for this investment is closest to
a. 8%   c. 12%
b. 10%   d. 18%

A.66 An individual bought a one-year savings certificate for $10,000, and it pays 6%. He has a taxable income that puts him at the 28% incremental income tax rate. His after-tax rate of return on this investment is closest to
a. 2%   c. 4%
b. 3%   d. 5%

A.67 A tool costing $300 has no salvage value. Its resulting before-tax cash flow is shown in the following partially completed cash flow table.

Year	Before-Tax Cash Flow	Effect on SOYD Deprec	Effect on Taxable Income	Income Taxes	After-Tax Cash Flow
0	−$300				
1	+100				
2	+150				
3	+200				

The tool is to be depreciated over three years using sum-of-years-digits depreciation. The income tax rate is 50%. The after-tax rate of return is nearest to
a. 8%   c. 12%
b. 10%   d. 15%

A.68 An engineer is considering the purchase of an annuity that will pay $1000 per year for 10 years. The engineer feels he should obtain a 5% rate of return on the annuity after considering the effect of an estimated 6% inflation per year. The amount he would be willing to pay to purchase the annuity is closest to
a. $1500   c. $4500
b. $3000   d. $6000

A.69 An automobile costs $20,000 today. You can earn 12% tax free on an "auto purchase account." If you expect the cost of the auto to increase by 10% per year, the amount you would need to deposit in the account to provide for the purchase of the auto five years from now is closest to
a. $12,000   c. $16,000
b. $14,000   d. $18,000

A.70 An engineer purchases a building lot for $40,000 cash and plans to sell it after five years. If he wants an 18% before-tax rate of return, after taking the 6% annual inflation rate into account, the selling price must be nearest to
a. $100,000   c. $150,000
b. $125,000   d. $175,000

A.71   A piece of equipment with a list price of $450 can actually be purchased for either $400 cash or $50 immediately plus four additional annual payments of $115.25. All values are in dollars of current purchasing power. If the typical customer considered a 5% interest rate appropriate, the inflation rate at which the two purchase alternatives are equivalent is nearest to

a.   5%   c.   8%
b.   6%   d.   10%

A.72   A man wants to determine whether to invest $1000 in a friend's speculative venture. He will do so if he thinks he can get this money back. The probabilities of the various outcomes at the end of one year are:

Result	Probability
$2000 (double his money)	0.3
1500	0.1
1000	0.2
500	0.3
0 (lose everything)	0.1

His expected outcome if he invests the $1000 is closest to

a. $800   c. $1000
b. $900   d. $1100

A.73   The amount you would be willing to pay for an insurance policy protecting you against a one in twenty chance of losing $10,000 three years from now, if interest is 10%, is closest to

a. $175   c. $1000
b. $350   d. $1500

SOLUTIONS

A.1   **b.**

$$F = P + Pin$$
$$600 + P = P + P(0.08)(2.50)$$
$$P = [600]/[0.08(2.50)] = \$3000$$

A.2   **b.**

$$F = P + Pin$$
$$1700 = 1000 + 1000(0.10)(n)$$
$$n = [700]/[1000(0.10)] = 7 \text{ years}$$

A.3   **d.**

$$F = A(F/A, i, n) = 400(F/A, 2\%, 100)$$
$$= 400(312.33) = \$124,890$$

A.4   **a.**   The relationship between the capital recovery factor and the sinking fund factor is $(A/P,i,n) = (A/F,i,n) + i$. Substituting the values in the problem

$$0.2091 = 0.1941 + i$$
$$i = 0.2091 - 0.1941 = 0.015 = 1\frac{1}{2}\%$$

A.5   **d.**

$$\begin{aligned} P &= A(P/A,i,n) + G(P/G,i,n) \\ &= 120(P/A,4\%,5) + 30(P/G,4\%,5) \\ &= 120(4.452) + 30(8.555) = \$791 \end{aligned}$$

A.6   **d.**

A.7   **c.**

At $i = 1\%/\text{month}$: $F = 1000(1 + 0.01)^{12} = \$1126.83$

At $i = 12\%/\text{year}$: $F = 1000(1 + 0.12)^{1} = \$1120.00$

Saving in interest charges $= 1126.83 - 1120.00 = \$6.83$

A.8   **a.**

$$P = Fe^{-rn} = 10,000e^{-0.09(10)} = 4066$$

A.9   **c.**   The nominal interest rate is the annual interest rate ignoring the effect of any compounding. Nominal interest rate $= 1\frac{1}{2}\% \times 12 = 18\%$.

A.10   **d.**   The interest paid per year $= 330 \times 4 = 1320$. The nominal annual interest rate $= 1320/10,000 = 0.132 = 13.2\%$.

A.11   **d.**

$$i_{\text{eff}} = (1 + r/m)^{m} - 1 = (1 + 0.03)^{6} - 1 = 0.194 = 19.4\%$$

A.12   **d.**

$$\begin{aligned} i_{\text{eff}} &= (1 + r/m)^{m} - 1 \\ r/m &= (1 + i_{\text{eff}})^{1/m} - 1 = (1 + 0.1956)^{1/12} = 0.015 \\ r &= 0.015(m) = 0.015 \times 12 = 0.18 = 18\% \end{aligned}$$

A.13   **a.**

$$i_{\text{eff}} = 20.52/300 = 0.0684 = 6.84\%$$

A.14   **d.**

$$\begin{aligned} i_{\text{eff}} &= (1 + r/m)^{m} - 1 \\ 0.12 &= (1 + r/12)^{12} - 1 \\ (1.12)^{1/12} &= (1 + r/12) \\ 1.00949 &= (1 + r/12) \\ r &= 0.00949 \times 12 = 0.1138 = 11.38\% \end{aligned}$$

A.15 **c.**

$$i_{eff} = (1 + r/m)^m - 1 = (1 + 0.10/365)^{365} - 1 = 0.1052 = 10.52\%$$

A.16 **c.**

$$i_{eff} = e^r - 1$$

where r = nominal annual interest rate

$$i_{eff} = e^{0.10} - 1 = 0.10517 = 10.52\%$$

A.17 **d.** For 3 months: $i_{eff} = e^r - 1$; $0.055 = e^r - 1$
The rate per quarter year is $r = \ln(1.055) = 0.05354$; $r = 4 \times 0.05354$
$= 0.214 = 21.4\%$ per year.

A.18 **c.**

$$i_{eff} = e^r - 1 = e^{0.25} - 1$$

A.19 **c.**

$$i_{eff} = e^r - 1 = 0.25$$
$$e^r = 1.25$$
$$\ln(e^r) = \ln(1.25)$$
$$r = \ln(1.25)$$

A.20 **c.**

$$P = F(P/F,i,n) = 500(P/F,2\%,10) = 500(0.8203) = \$410$$

A.21 **a.**

$$P = 300(P/A,8\%,20) = 300(9.818) = \$2945$$

A.22 **c.** Using single payment present worth factors:

$$P = 20(P/F,6\%,2) + 40(P/F,6\%,3) + 60(P/F,6\%,4)$$
$$+ 80(P/F,6\%,5) + 100(P/F,6\%,6) = \$229$$

Alternate solution using the gradient present worth factor:

$$P = 20(P/G,6\%,6) = 20(11.459) = \$229$$

A.23 **c.**

$$PW = 30(P/A,4\%,40) + 1000(P/F,4\%,40)$$
$$= 30(19.793) + 1000(0.2083) = \$802$$

A.24 **d.** Benefits are $1500 per year for the first five years and $1000 per year for the subsequent five years.

As Exhibit A.24 indicates, the benefits may be considered as $1000 per year for ten years, plus an additional $500 benefit in each of the first five years.

$$\text{Maximum investment} = \text{Present worth of benefits}$$
$$= 1000(P/A, 4\%, 10) + 500(P/A, 4\%, 5)$$
$$= 1000(8.111) + 500(4.452) = \$10,337$$

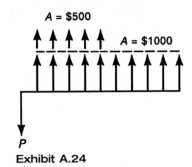

Exhibit A.24

A.25 c.

$$\text{NPW} = \text{PW of benefits} - \text{PW of cost}$$
$$= 2400(P/A,12\%,10) + 3000(P/F,12\%,10) - 10,000 = \$4526$$

A.26 d.

$$\text{PW of benefits} = \text{PW of cost}$$
$$5000 = 167.10(P/A,1.5\%,n)$$
$$(P/A,1.5\%,n) = 5000/167.10 = 29.92$$

From the $1\frac{1}{2}\%$ interest table, $n = 40$.

A.27 c.

$$\text{NPW} = \text{PW of benefts} - \text{PW of cost} = 0$$
$$= (650 - 300)(P/A,10\%,8) + S_8(P/F,10\%,8) - 2000 = 0$$
$$S_8 = 132.75/0.4665 = \$285$$

A.28 a. The amount of money needed now to begin the perpetual payments is $P' = A/i = 10,000/0.08 = 125,000$. From this we can compute the amount of money, P, that would need to have been deposited 50 years ago:

$$P = 125,000(P/F,8\%,50) = 125,000(0.0213) = \$2663$$

A.29 d.

$$P = A/i = 12,000/0.05 = \$240,000$$

A.30 b.

$$2 = 1(F/P,i,n)$$
$$(F/P, 2\%, n) = 2$$

From the 2% interest table, $n = $ about 35 months.

A.31 c. At end of 50 months

$$F = 10,000(F/P, 1\%, 50) = 10,000(1.645) = \$16,450$$

At end of 80 months

$$F = 16,450(F/P, 2\%, 10) = 16,450(1.219) = \$20,053$$

A.32 **d.**

$$FW = 200(F/A,1.5\%,12)(F/P,1.5\%,20)$$
$$= 200(13.041)(1.347) = \$3513$$

A.33 **d.** The present sum $P = 60$ is equivalent to Q six years hence at 6% interest. The future sum F may be calculated by either of two methods:

$$F = Q(F/P,6\%,4) \quad \text{and} \quad Q = 60(F/P,6\%,6)$$
$$F = P(F/P,6\%,10)$$

Since P is known, the second equation may be solved directly.

$$F = P(F/P,6\%,10) = 60(1.791) = \$107$$

A.34 **c.**

$$F' = A(F/A, i, n) = 200(F/A, 6\%, 15) = 200(23.276) = \$4655.20$$
$$F = F'(F/P, i, n) = 4655.20(F/P, 6\%, 1) = 4655.20(1.06) = \$4935$$

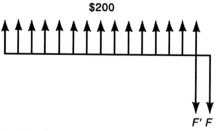

$200

F' F

Exhibit A.34

A.35 **c.**

$$EUAC = 150 + 150(A/G,8\%,4) = 150 + 150(1.404) = \$361$$

A.36 **b.**

$$\text{Annual payment} = 12,000(P/F,10\%,6)(A/P,10\%,8)$$
$$= 12,000(0.5645)(0.1874) = \$1269$$

A.37 **c.**

$$EUAC = 15,000(A/P,12\%,6) - 2000(A/F,12\%,6)$$
$$= 15,000(0.2432) - 2000(0.1232) = \$3402$$

A.38 **d.**

$$EUAC = 80,000(A/P,10\%,20) - 20,000(A/F,10\%,20)$$
$$+ \text{annual operating cost}$$
$$= 80,000\ (0.1175) - 20,000\ (0.0175) + 18,000$$
$$= 9400 - 350 + 18,000 = \$27,050$$

A.39 **b.**

$$EUAC = EUAB$$
$$27,000 = 80,000(A/P,10\%,20) + 18,000 - S(A/F,10\%,20)$$
$$= 80,000(0.1175) + 18,000 - S(0.0175)$$
$$S = (27,400 - 27,000)/0.0175 = \$22,857$$

A.40 **d.** The general equation for an infinite life, $P = A/i$, must be used to solve the problem.

$$i_{\text{eff}} = (1 + 0.025)^2 - 1 = 0.050625$$

The maximum annual withdrawal will be $A = Pi = 25,000(0.050625) = \$1266$.

A.41 **c.**

$$\text{Minimum sale price} = 10,000(F/P, 10\%, 10) + 100(F/A, 10\%, 10)$$
$$= 10,000(2.594) + 100(15.937) = \$27,530$$

A.42 **c.**

$$\text{NPW} = 1000(P/A, i, 20) - 10,000 = 0$$
$$(P/A, i, 20) = 10,000/1000 = 10$$

From interest tables: $6\% < i < 8\%$.

A.43 **a.** The rate of return was $= 600/10,000 = 0.06 = 6\%$.

A.44 **b.**

$$F = A(F/A, i, n)$$
$$12,000 = 1000(F/A, i, 10)$$
$$(F/A, i, 10) = 12,000/1000 = 12$$

In the 4% interest table: $(F/A, 4\%, 10) = 12.006$, so $i = 4\%$.

A.45 **b.**

$$\text{PW of cost} = \text{PW of benefits}$$
$$80,000 = (25,700 - 18,000)(P/A, i, 20) + (20,000(P/F, i, 20)$$

Try $i = 8\%$.

$$80,000 = 7709(9.818) + 20,000(0.2145) = 79,889$$

Therefore, the rate of return is very close to 8%.

A.46 **c.**

$$\text{PW of cost} = \text{PW of benefits}$$
$$100 = 102.15(P/F, i, 3) + 102.15(P/F, i, 6)$$

Solve by trial and error:
Try $i = 12\%$.
$$100 = 102.15(0.7118) + 102.15(0.5066) = 124.46$$

The PW of benefits exceeds the PW of cost. This indicates that the interest rate i is too low. Try $i = 18\%$.
$$100 = 102.15(0.6086) + 102.15(0.3704) = 100.00$$

Therefore, the rate of return is 18%.

A.47   **c.**   The difference between the alternatives:

Incremental cost $= 6000 - 2500 = \$3500$

Incremental annual benefit $= 1664 - 746 = \$918$

PW of cost = PW of benefits

$$3500 = 918(P/A, i, 5)$$
$$(P/A, i, 5) = 3500/918 = 3.81$$

From the interest tables, i is very close to 10%.

A.48   **c.**

$$B/C = \frac{\text{PW of benefits}}{\text{PW of cost}} = \frac{10,000(P/A, 12\%, 8) + 1000(P/G, 12\%, 8)}{50,000}$$

$$= \frac{10,000(4.968) + 1000(14.471)}{50,000} = 1.28$$

A.49   **c.**

$$B/C = \frac{\text{PW of benefits}}{\text{PW of cost}} = \frac{60(P/A, 12\%, 8)}{300} = \frac{60(4.968)}{300} = 0.99$$

Alternate solution:

$$B/C = \frac{\text{EUAB}}{\text{EUAC}} = \frac{60}{300(A/P, 12\%, 8)} = \frac{60}{300(0.2013)} = 0.99$$

A.50   **a.**

$$B/C = \frac{\text{PW of benefits}}{\text{PW of cost}} = \frac{14,000(P/A, 18\%, 8)}{57,000} = \frac{14,000(4.078)}{57,000} = 1.00$$

A.51   **d.**

$$B/C = \frac{\text{EUAB}}{\text{EUAC}} = \frac{20,000,000 - 5,500,000}{60,000,000(A/P, 8\%, 10)} = 1.62$$

A.52   **c.**

$$F = 10,000(F/P, 6\%, 5) - 1000(F/A, 6\%, 5)$$
$$= 10,000(1.338) - 1000(5.637) = \$7743$$

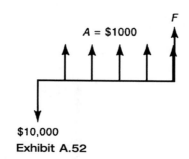

$A = \$1000$

$\$10,000$

Exhibit A.52

A.53 **d.**

$$\text{PW of cost}_A = \text{PW of cost}_B$$
$$55,000 + 16,200(P/A,10\%,n) = 75,000 + 12,450(P/A,10\%,n)$$
$$(P/A,10\%,n) = (75,000 - 55,000)/(16,200 - 12,450)$$
$$= 5.33$$

From the 10% interest table, $n = 8$ years.

A.54 **c.**

$$\text{EUAC}_{\text{untreated}} = \text{EUAC}_{\text{treated}}$$
$$350(A/P,6\%,6) = 700(A/P,6\%,n)$$
$$350(0.2034) = 700(A/P,6\%,n)$$
$$(A/P,6\%,n) = 71.19/700 = 0.1017$$

From the 6% interest table, $n = 15+$ years.

A.55 **d.**

$$\text{EUAC}_{\text{T-C}} = \text{EUAC}_{\text{Quick}}$$
$$(45+40)(A/P,12\%,n) = (22+40)(A/P,12\%,5)$$
$$(A/P,12\%,n) = 17.20/85 = 0.202$$

From the 12% interest table, $n = 8$.

A.56 **d.** At breakeven,

$$\text{NPW}_A = \text{NPW}_B$$
$$150(P/A,8\%,10) + 100(P/F,8\%,10) - 1000 = \text{NAB}(P/A,8\%,10)$$
$$+ 400(P/F,8\%,10) - 2000$$
$$52.82 = 6.71(\text{NAB}) - 1814.72$$

Net annual benefit (NAB) = (1814.72 + 52.82)/6.71 = $278.

A.57 **a.** "Double entry" probably is a reference to double entry accounting. It is not a method of depreciation.

A.58 **d.** Sum-of-years-digits depreciation:

$$D_j = \frac{n-j+1}{\frac{n}{2}(n+1)}(C - S_n)$$

$$D_1 = \frac{20-1+1}{\frac{20}{2}(20+1)}(80,000 - 20,000) = 5714$$

$$D_2 = \frac{20-2+1}{\frac{20}{2}(20+1)}(80,000 - 20,000) = 5429$$

Total: $\overline{\$11,143}$

Book value = Cost − Depreciation to date
= 80,000 − 11,143 = $68,857

A.59   **d.**   Double-declining-balance depreciation:

$$BV_j = C\left(1 - \frac{2}{n}\right)^j$$

$$BV_3 = 100,000\left(1 - \frac{2}{25}\right)^3 = \$77,869$$

A.60   **d.** $D_3 = (C - S)/n = (15,000 - 1000)/3 = \$4666$

A.61   **c.**   The half-year convention applies here; double-declining balance must be used with an assumed salvage value of zero. In general,

$$D_j = \frac{2C}{n}\left(1 - \frac{2}{n}\right)^{j-1}$$

For a half-year in Year 1:

$$D_1 = \frac{1}{2} \times \frac{2 \times 15,000}{3}\left(1 - \frac{2}{3}\right)^{1-1} = \$5000$$

A.62   **d.**

$$D_j = \frac{n - j + 1}{\frac{n}{2}(n+1)}(C - S_n)$$

$$D_2 = \frac{3 - 2 + 1}{\frac{3}{2}(3+1)}(15,000 - 1000) = \$4667$$

A.63   **d.**

$$D_2 = (15,000 - 1000)(A/F,8\%,3)(F/P,8\%,1)$$
$$= 14,000(0.3080)(1.08) = \$4657$$

A.64   **b.**   Twenty-eight percent of the $100 interest income must be paid in taxes. The balance of $72 is the after-tax income. Thus the after-tax rate of return = 72/1000 = 0.072 = 7.2%.

A.65   **d.**

Year	Before-Tax Cash Flow	SL Deprec	Taxable Income	34% Income Taxes	After-Tax Cash Flow
0	−$20,000				−$20,000
1–8	+6000	2500	3500	1190	+4810

$$D_j = (P - S)/n = \frac{20,000 - 0}{8} = 2500$$

PW of cost = PW of benefits
$$20,000 = 4810(P/A,i,8)$$
$$(P/A,i,8) = 20,000/4810 = 4.158$$

From interest tables, $i = 18\%$.

A.66 **c.** Additional taxable income = $0.06(10,000) = 600$

Additional income tax = $0.28(600) = 168$

After-tax rate income = $600 - 168 = 432$

After-tax rate of return = $432/10,000 = 0.043 = 4.3\%$

Alternate solution:

After-tax rate of return = $(1 - \text{Incremental tax rate})(\text{Before-tax rate of return})$
$$= (1 - 0.28)(0.06) = 0.043 = 4.3\%$$

A.67 **c.**

Year	Before-Tax Cash Flow	Effect on SOYD Deprec	Effect on Taxable Income	50% Income Taxes	After-Tax Cash Flow
0	−$300				−$300
1	+100	−150	−50	+25	+125
2	+150	−100	+50	−25	+125
3	+200	−50	+150	−75	+125

For the after-tax cash flow:

$$\text{PW of cost} = \text{PW of benefits}$$
$$300 = 125(P/A,i,3)$$
$$(P/A,i,3) = 300/125 = 2.40$$

From the interest table, we find that i (the after-tax rate of return) is close to 12.

A.68 **d.**

$$d = i + f + if = 0.05 + 0.06 + 0.05(0.06) = 0.113 = 11.3\%$$
$$P = A(P/A,11.3\%,10) = 100\left[\frac{(1+0.113)^{10} - 1}{0.113(1+0.113)^{10}}\right]$$
$$= 1000\left[\frac{1.9171}{0.3296}\right] = \$5816$$

A.69 **d.**

Cost of auto 5 years hence $(F) = P(1 + \text{inflation rate})^n$
$$= 20,000(1 + 0.10)^5 = 32,210$$

Amount to deposit now to have $32,210 available 5 years hence:

$$P = F(P/F,i,n) = 32,210(P/F,12\%,5) = 32,210(0.5674) = \$18,276$$

A.70 **b.**

Selling price $(F) = 40,000(F/P,18\%,5)(F/P,6\%,5)$
$$= 40,000(2.288)(1.338) = \$122,500$$

A.71   **b.**

$$PW \text{ of cash purchase} = PW \text{ of installment purchase}$$
$$400 = 50 + 115.25(P/A, d, 4)$$
$$(P/A, d, 4) = 350/115.25 = 3.037$$

From the interest tables, $d = 12\%$.

$$d = i + f + i(f)$$
$$0.12 = 0.05 + f + 0.05f$$
$$f = 0.07/1.05 = 0.0667 = 6.67\%$$

A.72   **d.**   The expected income $= 0.3(2000) + 0.1(1500) + 0.2(1000) + 0.3(500)$
$+ 0.1(0) = \$1100$.

A.73   **b.**

$$PW \text{ of benefit} = (10,000/20)(P/F, 10\%, 3)$$
$$= 500(0.7513) = \$376$$

Compound interest factors

| $\frac{1}{2}\%$ | | | | | | | | | $\frac{1}{2}\%$ |

	Single Payment		Uniform Payment Series				Uniform Gradient		
	Compound Amount Factor	Present Worth Factor	Sinking Fund Factor	Capital Recovery Factor	Compound Amount Factor	Present Worth Factor	Gradient Uniform Series	Gradient Present Worth	
n	Find F Given P F/P	Find P Given F P/F	Find A Given F A/F	Find A Given P A/P	Find F Given A F/A	Find P Given A P/A	Find A Given G A/G	Find P Given G P/G	n
1	1.005	.9950	1.0000	1.0050	1.000	0.995	0	0	1
2	1.010	.9901	.4988	.5038	2.005	1.985	0.499	0.991	2
3	1.015	.9851	.3317	.3367	3.015	2.970	0.996	2.959	3
4	1.020	.9802	.2481	.2531	4.030	3.951	1.494	5.903	4
5	1.025	.9754	.1980	.2030	5.050	4.926	1.990	9.803	5
6	1.030	.9705	.1646	.1696	6.076	5.896	2.486	14.660	6
7	1.036	.9657	.1407	.1457	7.106	6.862	2.980	20.448	7
8	1.041	.9609	.1228	.1278	8.141	7.823	3.474	27.178	8
9	1.046	.9561	.1089	.1139	9.182	8.779	3.967	34.825	9
10	1.051	.9513	.0978	.1028	10.228	9.730	4.459	43.389	10
11	1.056	.9466	.0887	.0937	11.279	10.677	4.950	52.855	11
12	1.062	.9419	.0811	.0861	12.336	11.619	5.441	63.218	12
13	1.067	.9372	.0746	.0796	13.397	12.556	5.931	74.465	13
14	1.072	.9326	.0691	.0741	14.464	13.489	6.419	86.590	14
15	1.078	.9279	.0644	.0694	15.537	14.417	6.907	99.574	15
16	1.083	.9233	.0602	.0652	16.614	15.340	7.394	113.427	16
17	1.088	.9187	.0565	.0615	17.697	16.259	7.880	128.125	17
18	1.094	.9141	.0532	.0582	18.786	17.173	8.366	143.668	18
19	1.099	.9096	.0503	.0553	19.880	18.082	8.850	160.037	19
20	1.105	9051	.0477	.0527	20.979	18.987	9.334	177.237	20
21	1.110	.9006	.0453	.0503	22.084	19.888	9.817	195.245	21
22	1.116	.8961	.0431	.0481	23.194	20.784	10.300	214.070	22
23	1.122	.8916	.0411	.0461	24.310	21.676	10.781	233.680	23
24	1.127	.8872	.0393	.0443	25.432	22.563	11.261	254.088	24
25	1.133	.8828	.0377	.0427	26.559	23.446	11.741	275.273	25
26	1.138	.8784	.0361	.0411	27.692	24.324	12.220	297.233	26
27	1.144	.8740	.0347	.0397	28.830	25.198	12.698	319.955	27
28	1.150	.8697	.0334	.0384	29.975	26.068	13.175	343.439	28
29	1.156	.8653	.0321	.0371	31.124	26.933	13.651	367.672	29
30	1.161	.8610	.0310	.0360	32.280	27.794	14.127	392.640	30
36	1.197	.8356	.0254	.0304	39.336	32.871	16.962	557.564	36
40	1.221	.8191	.0226	.0276	44.159	36.172	18.836	681.341	40
48	1.270	.7871	.0185	.0235	54.098	42.580	22.544	959.928	48
50	1.283	.7793	.0177	.0227	56.645	44.143	23.463	1 035.70	50
52	1.296	.7716	.0169	.0219	59.218	45.690	24.378	1 113.82	52
60	1.349	.7414	.0143	.0193	69.770	51.726	28.007	1 448.65	60
70	1.418	.7053	.0120	.0170	83.566	58.939	32.468	1 913.65	70
72	1.432	.6983	.0116	.0166	86.409	60.340	33.351	2 012.35	72
80	1.490	.6710	.0102	.0152	98.068	65.802	36.848	2 424.65	80
84	1.520	.6577	.00961	.0146	104.074	68.453	38.576	2 640.67	84
90	1.567	.6383	.00883	.0138	113.311	72.331	41.145	2 976.08	90
96	1.614	.6195	.00814	.0131	122.829	76.095	43.685	3 324.19	96
100	1.647	.6073	.00773	.0127	129.334	78.543	45.361	3 562.80	100
104	1.680	.5953	.00735	.0124	135.970	80.942	47.025	3 806.29	104
120	1.819	.5496	.00610	.0111	163.880	90.074	53.551	4 823.52	120
240	3.310	.3021	.00216	.00716	462.041	139.581	96.113	13 415.56	240
360	6.023	.1660	.00100	.00600	1 004.5	166.792	128.324	21 403.32	360
480	10.957	.0913	.00050	.00550	1 991.5	181.748	151.795	27 588.37	480

Compound interest factors

	Single Payment		Uniform Payment Series				Uniform Gradient		
	Compound Amount Factor	Present Worth Factor	Sinking Fund Factor	Capital Recovery Factor	Compound Amount Factor	Present Worth Factor	Gradient Uniform Series	Gradient Present Worth	
n	Find F Given P F/P	Find P Given F P/F	Find A Given F A/F	Find A Given P A/P	Find F Given A F/A	Find P Given A P/A	Find A Given G A/G	Find P Given G P/G	n
1	1.010	.9901	1.0000	1.0100	1.000	0.990	0	0	1
2	1.020	.9803	.4975	.5075	2.010	1.970	0.498	0.980	2
3	1.030	.9706	.3300	.3400	3.030	2.941	0.993	2.921	3
4	1.041	.9610	.2463	.2563	4.060	3.902	1.488	5.804	4
5	1.051	.9515	.1960	.2060	5.101	4.853	1.980	9.610	5
6	1.062	.9420	.1625	.1725	6.152	5.795	2.471	14.320	6
7	1.072	.9327	.1386	.1486	7.214	6.728	2.960	19.917	7
8	1.083	.9235	.1207	.1307	8.286	7.652	3.448	26.381	8
9	1.094	.9143	.1067	.1167	9.369	8.566	3.934	33.695	9
10	1.105	.9053	.0956	.1056	10.462	9.471	4.418	41.843	10
11	1.116	.8963	.0865	.0965	11.567	10.368	4.900	50.806	11
12	1.127	.8874	.0788	.0888	12.682	11.255	5.381	60.568	12
13	1.138	.8787	.0724	.0824	13.809	12.134	5.861	71.112	13
14	1.149	.8700	.0669	.0769	14.947	13.004	6.338	82.422	14
15	1.161	.8613	.0621	.0721	16.097	13.865	6.814	94.481	15
16	1.173	.8528	.0579	.0679	17.258	14.718	7.289	107.273	16
17	1.184	.8444	.0543	.0643	18.430	15.562	7.761	120.783	17
18	1.196	.8360	.0510	.0610	19.615	16.398	8.232	134.995	18
19	1.208	.8277	.0481	.0581	20.811	17.226	8.702	149.895	19
20	1.220	.8195	.0454	.0554	22.019	18.046	9.169	165.465	20
21	1.232	.8114	.0430	.0530	23.239	18.857	9.635	181.694	21
22	1.245	.8034	.0409	.0509	24.472	19.660	10.100	198.565	22
23	1.257	.7954	.0389	.0489	25.716	20.456	10.563	216.065	23
24	1.270	.7876	.0371	.0471	26.973	21.243	11.024	234.179	24
25	1.282	.7798	.0354	.0454	28.243	22.023	11.483	252.892	25
26	1.295	.7720	0339	.0439	29.526	22.795	11.941	272.195	26
27	1.308	.7644	.0324	.0424	30.821	23.560	12.397	292.069	27
28	1.321	.7568	.0311	.0411	32.129	24.316	12.852	312.504	28
29	1.335	.7493	.0299	.0399	33.450	25.066	13.304	333.486	29
30	1.348	.7419	.0287	.0387	34.785	25.808	13.756	355.001	30
36	1.431	.6989	.0232	.0332	43.077	30.107	16.428	494.620	36
40	1.489	.6717	.0205	.0305	48.886	32.835	18.178	596.854	40
48	1.612	.6203	.0163	.0263	61.223	37.974	21.598	820.144	48
50	1.645	.6080	.0155	.0255	64.463	39.196	22.436	879.417	50
52	1.678	.5961	.0148	.0248	67.769	40.394	23.269	939.916	52
60	1.817	.5504	.0122	.0222	81.670	44.955	26.533	1 192.80	60
70	2.007	.4983	.00993	.0199	100.676	50.168	30.470	1 528.64	70
72	2.047	.4885	.00955	.0196	104.710	51.150	31.239	1 597.86	72
80	2.217	.4511	.00822	.0182	121.671	54.888	34.249	1 879.87	80
84	2.307	.4335	.00765	.0177	130.672	56.648	35.717	2 023.31	84
90	2.449	.4084	.00690	.0169	144.863	59.161	37.872	2 240.56	90
96	2.599	.3847	.00625	.0163	159.927	61.528	39.973	2 459.42	96
100	2.705	.3697	.00587	.0159	170.481	63.029	41.343	2 605.77	100
104	2.815	.3553	.00551	.0155	181.464	64.471	42.688	2 752.17	104
120	3.300	.3030	.00435	.0143	230.039	69.701	47.835	3 334.11	120
240	10.893	.0918	.00101	.0110	989.254	90.819	75.739	6 878.59	240
360	35.950	.0278	.00029	.0103	3 495.0	97.218	89.699	8 720.43	360
480	118.648	.00843	.00008	.0101	11 764.8	99.157	95.920	9 511.15	480

Compound interest factors

	Single Payment		Uniform Payment Series				Uniform Gradient		
	Compound Amount Factor	Present Worth Factor	Sinking Fund Factor	Capital Recovery Factor	Compound Amount Factor	Present Worth Factor	Gradient Uniform Series	Gradient Present Worth	
	Find F Given P	Find P Given F	Find A Given F	Find A Given P	Find F Given A	Find P Given A	Find A Given G	Find P Given G	
n	F/P	P/F	A/F	A/P	F/A	P/A	A/G	P/G	n
1	1.015	.9852	1.0000	1.0150	1.000	0.985	0	0	1
2	1.030	.9707	.4963	.5113	2.015	1.956	0.496	0.970	2
3	1.046	.9563	.3284	.3434	3.045	2.912	0.990	2.883	3
4	1.061	.9422	.2444	.2594	4.091	3.854	1.481	5.709	4
5	1.077	.9283	.1941	.2091	5.152	4.783	1.970	9.422	5
6	1.093	.9145	.1605	.1755	6.230	5.697	2.456	13.994	6
7	1.110	.9010	.1366	.1516	7.323	6.598	2.940	19.400	7
8	1.126	.8877	.1186	.1336	8.433	7.486	3.422	25.614	8
9	1.143	.8746	.1046	.1196	9.559	8.360	3.901	32.610	9
10	1.161	.8617	.0934	.1084	10.703	9.222	4.377	40.365	10
11	1.178	.8489	.0843	.0993	11.863	10.071	4.851	48.855	11
12	1.196	.8364	.0767	.0917	13.041	10.907	5.322	58.054	12
13	1.214	.8240	.0702	.0852	14.237	11.731	5.791	67.943	13
14	1.232	.8118	.0647	.0797	15.450	12.543	6.258	78.496	14
15	1.250	.7999	.0599	.0749	16.682	13.343	6.722	89.694	15
16	1.269	.7880	.0558	.0708	17.932	14.131	7.184	101.514	16
17	1.288	.7764	.0521	.0671	19.201	14.908	7.643	113.937	17
18	1.307	.7649	.0488	.0638	20.489	15.673	8.100	126.940	18
19	1.327	.7536	.0459	.0609	21.797	16.426	8.554	140.505	19
20	1.347	.7425	.0432	.0582	23.124	17.169	9.005	154.611	20
21	1.367	.7315	.0409	.0559	24.470	17.900	9.455	169.241	21
22	1.388	.7207	.0387	.0537	25.837	18.621	9.902	184.375	22
23	1.408	.7100	.0367	.0517	27.225	19.331	10.346	199.996	23
24	1.430	.6995	.0349	.0499	28.633	20.030	10.788	216.085	24
25	1.451	.6892	.0333	.0483	30.063	20.720	11.227	232.626	25
26	1.473	.6790	.0317	.0467	31.514	21.399	11.664	249.601	26
27	1.495	.6690	.0303	.0453	32.987	22.068	12.099	266.995	27
28	1.517	.6591	.0290	.0440	34.481	22.727	12.531	284.790	28
29	1.540	.6494	.0278	.0428	35.999	23.376	12.961	302.972	29
30	1.563	.6398	.0266	.0416	37.539	24.016	13.388	321.525	30
36	1.709	.5851	.0212	.0362	47.276	27.661	15.901	439.823	36
40	1.814	.5513	.0184	.0334	54.268	29.916	17.528	524.349	40
48	2.043	.4894	.0144	.0294	69.565	34.042	20.666	703.537	48
50	2.105	.4750	.0136	.0286	73.682	35.000	21.428	749.955	50
52	2.169	.4611	.0128	.0278	77.925	35.929	22.179	796.868	52
60	2.443	.4093	.0104	.0254	96.214	39.380	25.093	988.157	60
70	2.835	.3527	.00817	.0232	122.363	43.155	28.529	1 231.15	70
72	2.921	.3423	.00781	.0228	128.076	43.845	29.189	1 279.78	72
80	3.291	.3039	.00655	.0215	152.710	46.407	31.742	1 473.06	80
84	3.493	.2863	.00602	.0210	166.172	47.579	32.967	1 568.50	84
90	3.819	.2619	.00532	.0203	187.929	49.210	34.740	1 709.53	90
96	4.176	.2395	.00472	.0197	211.719	50.702	36.438	1 847.46	96
100	4.432	.2256	.00437	.0194	228.802	51.625	37.529	1 937.43	100
104	4.704	.2126	.00405	.0190	246.932	52.494	38.589	2 025.69	104
120	5.969	.1675	.00302	.0180	331.286	55.498	42.518	2 359.69	120
240	35.632	.0281	.00043	.0154	2 308.8	64.796	59.737	3 870.68	240
360	212.700	.00470	.00007	.0151	14 113.3	66.353	64.966	4 310.71	360
480	1 269.7	.00079	.00001	.0150	84 577.8	66.614	66.288	4 415.74	480

Compound interest factors

2%   **2%**

	Single Payment		Uniform Payment Series				Uniform Gradient		
	Compound Amount Factor	Present Worth Factor	Sinking Fund Factor	Capital Recovery Factor	Compound Amount Factor	Present Worth Factor	Gradient Uniform Series	Gradient Present Worth	
	Find F Given P	Find P Given F	Find A Given F	Find A Given P	Find F Given A	Find P Given A	Find A Given G	Find P Given G	
n	F/P	P/F	A/F	A/P	F/A	P/A	A/G	P/G	n
1	1.020	.9804	1.0000	1.0200	1.000	0.980	0	0	1
2	1.040	.9612	.4951	.5151	2.020	1.942	0.495	0.961	2
3	1.061	.9423	.3268	.3468	3.060	2.884	0.987	2.846	3
4	1.082	.9238	.2426	.2626	4.122	3.808	1.475	5.617	4
5	1.104	.9057	.1922	.2122	5.204	4.713	1.960	9.240	5
6	1.126	.8880	.1585	.1785	6.308	5.601	2.442	13.679	6
7	1.149	.8706	.1345	.1545	7.434	6.472	2.921	18.903	7
8	1.172	.8535	.1165	.1365	8.583	7.325	3.396	24.877	8
9	1.195	.8368	.1025	.1225	9.755	8.162	3.868	31.571	9
10	1.219	.8203	.0913	.1113	10.950	8.983	4.337	38.954	10
11	1.243	.8043	.0822	.1022	12.169	9.787	4.802	46.996	11
12	1.268	.7885	.0746	.0946	13.412	10.575	5.264	55.669	12
13	1.294	.7730	.0681	.0881	14.680	11.348	5.723	64.946	13
14	1.319	.7579	.0626	.0826	15.974	12.106	6.178	74.798	14
15	1.346	.7430	.0578	.0778	17.293	12.849	6.631	85.200	15
16	1.373	.7284	.0537	.0737	18.639	13.578	7.080	96.127	16
17	1.400	.7142	.0500	.0700	20.012	14.292	7.526	107.553	17
18	1.428	.7002	.0467	.0667	21.412	14.992	7.968	119.456	18
19	1.457	.6864	.0438	.0638	22.840	15.678	8.407	131.812	19
20	1.486	.6730	.0412	.0612	24.297	16.351	8.843	144.598	20
21	1.516	.6598	.0388	.0588	25.783	17.011	9.276	157.793	21
22	1.546	.6468	.0366	.0566	27.299	17.658	9.705	171.377	22
23	1.577	.6342	.0347	.0547	28.845	18.292	10.132	185.328	23
24	1.608	.6217	.0329	.0529	30.422	18.914	10.555	199.628	24
25	1.641	.6095	.0312	.0512	32.030	19.523	10.974	214.256	25
26	1.673	.5976	.0297	.0497	33.671	20.121	11.391	229.169	26
27	1.707	.5859	.0283	.0483	35.344	20.707	11.804	244.428	27
28	1.741	.5744	.0270	.0470	37.051	21.281	12.214	259.936	28
29	1.776	.5631	.0258	.0458	38.792	21.844	12.621	275.703	29
30	1.811	.5521	.0247	.0447	40.568	22.396	13.025	291.713	30
36	2.040	.4902	.0192	.0392	51.994	25.489	15.381	392.036	36
40	2.208	.4529	.0166	.0366	60.402	27.355	16.888	461.989	40
48	2.587	.3865	.0126	.0326	79.353	30.673	19.755	605.961	48
50	2.692	.3715	.0118	.0318	84.579	31.424	20.442	642.355	50
52	2.800	.3571	.0111	.0311	90.016	32.145	21.116	678.779	52
60	3.281	.3048	.00877	.0288	114.051	34.761	23.696	823.692	60
70	4.000	.2500	.00667	.0267	149.977	37.499	26.663	999.829	70
72	4.161	.2403	.00633	.0263	158.056	37.984	27.223	1 034.050	72
80	4.875	.2051	.00516	.0252	193.771	39.744	29.357	1 166.781	80
84	5.277	.1895	.00468	.0247	213.865	40.525	30.361	1 230.413	84
90	5.943	.1683	.00405	.0240	247.155	41.587	31.793	1 322.164	90
96	6.693	.1494	.00351	.0235	284.645	42.529	33.137	1 409.291	96
100	7.245	.1380	.00320	.0232	312.230	43.098	33.986	1 464.747	100
104	7.842	.1275	.00292	.0229	342.090	43.624	34.799	1 518.082	104
120	10.765	.0929	.00205	.0220	488.255	45.355	37.711	1 710.411	120
240	115.887	.00863	.00017	.0202	5 744.4	49.569	47.911	2 374.878	240
360	1 247.5	.00080	.00002	.0200	62 326.8	49.960	49.711	2 483.567	360
480	13 429.8	.00007		.0200	671 442.0	49.996	49.964	2 498.027	480

Compound interest factors

4%									4%
	Single Payment		Uniform Payment Series				Uniform Gradient		
	Compound Amount Factor	Present Worth Factor	Sinking Fund Factor	Capital Recovery Factor	Compound Amount Factor	Present Worth Factor	Gradient Uniform Series	Gradient Present Worth	
n	Find F Given P F/P	Find P Given F P/F	Find A Given F A/F	Find A Given P A/P	Find F Given A F/A	Find P Given A P/A	Find A Given G A/G	Find P Given G P/G	n
1	1.040	.9615	1.0000	1.0400	1.000	0.962	0	0	1
2	1.082	.9246	.4902	.5302	2.040	1.886	0.490	0.925	2
3	1.125	.8890	.3203	.3603	3.122	2.775	0.974	2.702	3
4	1.170	.8548	.2355	.2755	4.246	3.630	1.451	5.267	4
5	1.217	.8219	.1846	.2246	5.416	4.452	1.922	8.555	5
6	1.265	.7903	.1508	.1908	6.633	5.242	2.386	12.506	6
7	1.316	.7599	.1266	.1666	7.898	6.002	2.843	17.066	7
8	1.369	.7307	.1085	.1485	9.214	6.733	3.294	22.180	8
9	1.423	.7026	.0945	.1345	10.583	7.435	3.739	27.801	9
10	1.480	.6756	.0833	.1233	12.006	8.111	4.177	33.881	10
11	1.539	.6496	.0741	.1141	13.486	8.760	4.609	40.377	11
12	1.601	.6246	.0666	.1066	15.026	9.385	5.034	47.248	12
13	1.665	.6006	.0601	.1001	16.627	9.986	5.453	54.454	13
14	1.732	.5775	.0547	.0947	18.292	10.563	5.866	61.962	14
15	1.801	.5553	.0499	.0899	20.024	11.118	6.272	69.735	15
16	1.873	.5339	.0458	.0858	21.825	11.652	6.672	77.744	16
17	1.948	.5134	.0422	.0822	23.697	12.166	7.066	85.958	17
18	2.029	.4936	.0390	.0790	25.645	12.659	7.453	94.350	18
19	2.107	.4746	.0361	.0761	27.671	13.134	7.834	102.893	19
20	2.191	.4564	.0336	.0736	29.778	13.590	8.209	111.564	20
21	2.279	.4388	.0313	.0713	31.969	14.029	8.578	120.341	21
22	2.370	.4220	.0292	.0692	34.248	14.451	8.941	129.202	22
23	2.465	.4057	.0273	.0673	36.618	14.857	9.297	138.128	23
24	2.563	.3901	.0256	.0656	39.083	15.247	9.648	147.101	24
25	2.666	.3751	.0240	.0640	41.646	15.622	9.993	156.104	25
26	2.772	.3607	.0226	.0626	44.312	15.983	10.331	165.121	26
27	2.883	.3468	.0212	.0612	47.084	16.330	10.664	174.138	27
28	2.999	.3335	.0200	.0600	49.968	16.663	10.991	183.142	28
29	3.119	.3207	.0189	.0589	52.966	16.984	11.312	192.120	29
30	3.243	.3083	.0178	.0578	56.085	17.292	11.627	201.062	30
31	3.373	.2965	.0169	.0569	59.328	17.588	11.937	209.955	31
32	3.508	.2851	.0159	.0559	62.701	17.874	12.241	218.792	32
33	3.648	.2741	.0151	.0551	66.209	18.148	12.540	227.563	33
34	3.794	.2636	.0143	.0543	69.858	18.411	12.832	236.260	34
35	3.946	.2534	.0136	.0536	73.652	18.665	13.120	244.876	35
40	4.801	.2083	.0105	.0505	95.025	19.793	14.476	286.530	40
45	5.841	.1712	.00826	.0483	121.029	20.720	15.705	325.402	45
50	7.107	.1407	.00655	.0466	152.667	21.482	16.812	361.163	50
55	8.646	.1157	.00523	.0452	191.159	22.109	17.807	393.689	55
60	10.520	.0951	.00420	.0442	237.990	22.623	18.697	422.996	60
65	12.799	.0781	.00339	.0434	294.968	23.047	19.491	449.201	65
70	15.572	.0642	.00275	.0427	364.290	23.395	20.196	472.479	70
75	18.945	.0528	.00223	.0422	448.630	23.680	20.821	493.041	75
80	23.050	.0434	.00181	.0418	551.244	23.915	21.372	511.116	80
85	28.044	.0357	.00148	.0415	676.089	24.109	21.857	526.938	85
90	34.119	.0293	.00121	.0412	827.981	24.267	22.283	540.737	90
95	41.511	.0241	.00099	.0410	1 012.8	24.398	22.655	552.730	95
100	50.505	.0198	.00081	.0408	1 237.6	24.505	22.980	563.125	100

Compound interest factors

6%									6%
	Single Payment		**Uniform Payment Series**				**Uniform Gradient**		
	Compound Amount Factor	Present Worth Factor	Sinking Fund Factor	Capital Recovery Factor	Compound Amount Factor	Present Worth Factor	Gradient Uniform Series	Gradient Present Worth	
	Find F Given P	Find P Given F	Find A Given F	Find A Given P	Find F Given A	Find P Given A	Find A Given G	Find P Given G	
n	F/P	P/F	A/F	A/P	F/A	P/A	A/G	P/G	n
1	1.060	.943	1.0000	1.0600	1.000	0.943	0	0	1
2	1.124	.8900	.4854	.5454	2.060	1.833	0.485	0.890	2
3	1.191	.8396	.3141	.3741	3.184	2.673	0.961	2.569	3
4	1.262	.7921	.2286	.2886	4.375	3.465	1.427	4.945	4
5	1.338	.7473	.1774	.2374	5.637	4.212	1.884	7.934	5
6	1.419	.7050	.1434	.2034	6.975	4.917	2.330	11.459	6
7	1.504	.6651	.1191	.1791	8.394	5.582	2.768	15.450	7
8	1.594	.6274	.1010	.1610	9.897	6.210	3.195	19.841	8
9	1.689	.5919	.0870	.1470	11.491	6.802	3.613	24.577	9
10	1.791	.5584	.0759	.1359	13.181	7.360	4.022	29.602	10
11	1.898	.5268	.0668	.1268	14.972	7.887	4.421	34.870	11
12	2.012	.4970	.0593	.1193	16.870	8.384	4.811	40.337	12
13	2.133	.4688	.0530	.1130	18.882	8.853	5.192	45.963	13
14	2.261	.4423	.0476	.1076	21.015	9.295	5.564	51.713	14
15	2.397	.4173	.0430	.1030	23.276	9.712	5.926	57.554	15
16	2.540	.3936	.0390	.0990	25.672	10.106	6.279	63.459	16
17	2.693	.3714	.0354	.0954	28.213	10.477	6.624	69.401	17
18	2.854	.3503	.0324	.0924	30.906	10.828	6.960	75.357	18
19	3.026	.3305	.0296	.0896	33.760	11.158	7.287	81.306	19
20	3.207	.3118	.0272	.0872	36.786	11.470	7.605	87.230	20
21	3.400	.2942	.0250	.0850	39.993	11.764	7.915	93.113	21
22	3.604	.2775	.0230	.0830	43.392	12.042	8.217	98.941	22
23	3.820	.2618	.0213	.0813	46.996	12.303	8.510	104.700	23
24	4.049	.2470	.0197	.0797	50.815	12.550	8.795	110.381	24
25	4.292	.2330	.0182	.0782	54.864	12.783	9.072	115.973	25
26	4.549	.2198	.0169	.0769	59.156	13.003	9.341	121.468	26
27	4.822	.2074	.0157	.0757	63.706	13.211	9.603	126.860	27
28	5.112	.1956	.0146	.0746	68.528	13.406	9.857	132.142	28
29	5.418	.1846	.0136	.0736	73.640	13.591	10.103	137.309	29
30	5.743	.1741	.0126	.0726	79.058	13.765	10.342	142.359	30
31	6.088	.1643	.0118	.0718	84.801	13.929	10.574	147.286	31
32	6.453	.1550	.0110	.0710	90.890	14.084	10.799	152.090	32
33	6.841	.1462	.0103	.0703	97.343	14.230	11.017	156.768	33
34	7.251	.1379	.00960	.0696	104.184	14.368	11.228	161.319	34
35	7.686	.1301	.00897	.0690	111.435	11.498	11.432	165.743	35
40	10.286	.0972	.00646	.0665	154.762	15.046	12.359	185.957	40
45	13.765	.0727	.00470	.0647	212.743	15.456	13.141	203.109	45
50	18.420	.0543	.00344	.0634	290.335	15.762	13.796	217.457	50
55	24.650	.0406	.00254	.0625	394.171	15.991	14.341	229.322	55
60	32.988	.0303	.00188	.0619	533.126	16.161	14.791	239.043	60
65	44.145	.0227	.00139	.0614	719.080	16.289	15.160	246.945	65
70	59.076	.0169	.00103	.0610	967.928	16.385	15.461	253.327	70
75	79.057	.0126	.00077	.0608	1 300.9	16.456	15.706	258.453	75
80	105.796	.00945	.00057	.0606	1 746.6	16.509	15.903	262.549	80
85	141.578	.00706	.00043	.0604	2 343.0	16.549	16.062	265.810	85
90	189.464	.00528	.00032	.0603	3 141.1	16.579	16.189	268.395	90
95	253.545	.00394	.00024	.0602	4 209.1	16.601	16.290	270.437	95
100	339.300	.00295	.00018	.0602	5 638.3	16.618	16.371	272.047	100

Compound interest factors

8%									8%
	Single Payment		**Uniform Payment Series**				**Uniform Gradient**		
	Compound Amount Factor	Present Worth Factor	Sinking Fund Factor	Capital Recovery Factor	Compound Amount Factor	Present Worth Factor	Gradient Uniform Series	Gradient Present Worth	
	Find F Given P	Find P Given F	Find A Given F	Find A Given P	Find F Given A	Find P Given A	Find A Given G	Find P Given G	
n	F/P	P/F	A/F	A/P	F/A	P/A	A/G	P/G	n
1	1.080	.9259	1.0000	1.0800	1.000	0.926	0	0	1
2	1.166	.8573	.4808	.5608	2.080	1.783	0.481	0.857	2
3	1.260	.7938	.3080	.3880	3.246	2.577	0.949	2.445	3
4	1.360	.7350	.2219	.3019	4.506	3.312	1.404	4.650	4
5	1.469	.6806	.1705	.2505	5.867	3.993	1.846	7.372	5
6	1.587	.6302	.1363	.2163	7.336	4.623	2.276	10.523	6
7	1.714	.5835	.1121	.1921	8.923	5.206	2.694	14.024	7
8	1.851	.5403	.0940	.1740	10.637	5.747	3.099	17.806	8
9	1.999	.5002	.0801	.1601	12.488	6.247	3.491	21.808	9
10	2.159	.4632	.0690	.1490	14.487	6.710	3.871	25.977	10
11	2.332	.4289	.0601	.1401	16.645	7.139	4.240	30.266	11
12	2.518	.3971	.0527	.1327	18.977	7.536	4.596	34.634	12
13	2.720	.3677	.0465	.1265	21.495	7.904	4.940	39.046	13
14	2.937	.3405	.0413	.1213	24.215	8.244	5.273	43.472	14
15	3.172	.3152	.0368	.1168	27.152	8.559	5.594	47.886	15
16	3.426	.2919	.0330	.1130	30.324	8.851	5.905	52.264	16
17	3.700	.2703	.0296	.1096	33.750	9.122	6.204	56.588	17
18	3.996	.2502	.0267	.1067	37.450	9.372	6.492	60.843	18
19	4.316	.2317	.0241	.1041	41.446	9.604	6.770	65.013	19
20	4.661	.2145	.0219	.1019	45.762	9.818	7.037	69.090	20
21	5.034	.1987	.0198	.0998	50.423	10.017	7.294	73.063	21
22	5.437	.1839	.0180	.0980	55.457	10.201	7.541	76.926	22
23	5.871	.1703	.0164	.0964	60.893	10.371	7.779	80.673	24
24	6.341	.1577	.0150	.0950	66.765	10.529	8.007	84.300	24
25	6.848	.1460	.0137	.0937	73.106	10.675	8.225	87.804	25
26	7.396	.1352	.0125	.0925	79.954	10.810	8.435	91.184	26
27	7.988	.1252	.0114	.0914	87.351	10.935	8.636	94.439	27
28	8.627	.1159	.0105	.0905	95.339	11.051	8.829	97.569	28
29	9.317	.1073	.00962	.0896	103.966	11.158	9.013	100.574	29
30	10.063	.0994	.00883	.0888	113.283	11.258	9.190	103.456	30
31	10.868	.0920	.00811	.0881	123.346	11.350	9.358	106.216	31
32	11.737	.0852	.00745	.0875	134.214	11.435	9.520	108.858	32
33	12.676	.0789	.00685	.0869	145.951	11.514	9.674	111.382	33
34	13.690	.0730	.00630	.0863	158.627	11.587	9.821	113.792	34
35	14.785	.0676	.00580	.0858	172.317	11.655	9.961	116.092	35
40	21.725	.0460	.00386	.0839	259.057	11.925	10.570	126.042	40
45	31.920	.0313	.00259	.0826	386.506	12.108	11.045	133.733	45
50	46.902	.0213	.00174	.0817	573.771	12.233	11.411	139.593	50
55	68.914	.0145	.00118	.0812	848.925	12.319	11.690	144.006	55
60	101.257	.00988	.00080	.0808	1 253.2	12.377	11.902	147.300	60
65	148.780	.00672	.00054	.0805	1 847.3	12.416	12.060	149.739	65
70	218.607	.00457	.00037	.0804	2 720.1	12.443	12.178	151.533	70
75	321.205	.00311	.00025	.0802	4 002.6	12.461	12.266	152.845	75
80	471.956	.00212	.00017	.0802	5 887.0	12.474	12.330	153.800	80
85	693.458	.00144	.00012	.0801	8 655.7	12.482	12.377	154.492	85
90	1 018.9	.00098	.00008	.0801	12 724.0	12.488	12.412	154.993	90
95	1 497.1	.00067	.00005	.0801	18 701.6	12.492	12.437	155.352	95
100	2 199.8	.00045	.00004	.0800	27 484.6	12.494	12.455	155.611	100

Compound interest factors

	Single Payment		Uniform Payment Series				Uniform Gradient		
	Compound Amount Factor	Present Worth Factor	Sinking Fund Factor	Capital Recovery Factor	Compound Amount Factor	Present Worth Factor	Gradient Uniform Series	Gradient Present Worth	
	Find F Given P	Find P Given F	Find A Given F	Find A Given P	Find F Given A	Find P Given A	Find A Given G	Find P Given G	
n	F/P	P/F	A/F	A/P	F/A	P/A	A/G	P/G	n
1	1.100	.9091	1.0000	1.1000	1.000	0.909	0	0	1
2	1.210	.8264	.4762	.5762	2.100	1.736	0.476	0.826	2
3	1.331	.7513	.3021	.4021	3.310	2.487	0.937	2.329	3
4	1.464	.6830	.2155	.3155	4.641	3.170	1.381	4.378	4
5	1.611	.6209	.1638	.2638	6.105	3.791	1.810	6.862	5
6	1.772	.5645	.1296	.2296	7.716	4.355	2.224	9.684	6
7	1.949	.5132	.1054	.2054	9.487	4.868	2.622	12.763	7
8	2.144	.4665	.0874	.1874	11.436	5.335	3.004	16.029	8
9	2.358	.4241	.0736	.1736	13.579	5.759	3.372	19.421	9
10	2.594	.3855	.0627	.1627	15.937	6.145	3.725	22.891	10
11	2.853	.3505	.0540	.1540	18.531	6.495	4.064	26.396	11
12	3.138	.3186	.0468	.1468	21.384	6.814	4.388	29.901	12
13	3.452	.2897	.0408	.1408	24.523	7.103	4.699	33.377	13
14	3.797	.2633	.0357	.1357	27.975	7.367	4.996	36.801	14
15	4.177	.2394	.0315	.1315	31.772	7.606	5.279	40.152	15
16	4.595	.2176	.0278	.1278	35.950	7.824	5.549	43.416	16
17	5.054	.1978	.0247	.1247	40.545	8.022	5.807	46.582	17
18	5.560	.1799	.0219	.1219	45.599	8.201	6.053	49.640	18
19	6.116	.1635	.0195	.1195	51.159	8.365	6.286	52.583	19
20	6.728	.1486	.0175	.1175	57.275	8.514	6.508	55.407	20
21	7.400	.1351	.0156	.1156	64.003	8.649	6.719	58.110	21
22	8.140	.1228	.0140	.1140	71.403	8.772	6.919	60.689	22
23	8.954	.1117	.0126	.1126	79.543	8.883	7.108	63.146	24
24	9.850	.1015	.0113	.1113	88.497	8.985	7.288	65.481	24
25	10.835	.0923	.0102	.1102	98.347	9.077	7.458	67.696	25
26	11.918	.0839	.00916	.1092	109.182	9.161	7.619	69.794	26
27	13.110	.0763	.00826	.1083	121.100	9.237	7.770	71.777	27
28	14.421	.0693	.00745	.1075	134.210	9.307	7.914	73.650	28
29	15.863	.0630	.00673	.1067	148.631	9.370	8.049	75.415	29
30	17.449	.0573	.00608	.1061	164.494	9.427	8.176	77.077	30
31	19.194	.0521	.00550	.1055	181.944	9.479	8.296	78.640	31
32	21.114	.0474	.00497	.1050	201.138	9.526	8.409	80.108	32
33	23.225	.0431	.00450	.1045	222.252	9.569	8.515	81.486	33
34	25.548	.0391	.00407	.1041	245.477	9.609	8.615	82.777	34
35	28.102	.0356	.00369	.1037	271.025	9.644	8.709	83.987	35
40	45.259	.0221	.00226	.1023	442.593	9.779	9.096	88.953	40
45	72.891	.0137	.00139	.1014	718.905	9.863	9.374	92.454	45
50	117.391	.00852	.00086	.1009	1 163.9	9.915	9.570	94.889	50
55	189.059	.00529	.00053	.1005	1 880.6	9.947	9.708	96.562	55
60	304.482	.00328	.00033	.1003	3 034.8	9.967	9.802	97.701	60
65	490.371	.00204	.00020	.1002	4 893.7	9.980	9.867	98.471	65
70	789.748	.00127	.00013	.1001	7 887.5	9.987	9.911	98.987	70
75	1 271.9	.00079	.00008	.1001	12 709.0	9.992	9.941	99.332	75
80	2 048.4	.00049	.00005	.1000	20 474.0	9.995	9.961	99.561	80
85	3 229.0	.00030	.00003	.1000	32 979.7	9.997	9.974	99.712	85
90	5 313.0	.00019	.00002	.1000	53 120.3	9.998	9.983	99.812	90
95	8 556.7	.00012	.00001	.1000	85 556.9	9.999	9.989	99.877	95
100	13 780.6	.00007	.00001	.1000	137 796.3	9.999	9.993	99.920	100

Compound interest factors

12%

	Single Payment		Uniform Payment Series				Uniform Gradient		
	Compound Amount Factor	Present Worth Factor	Sinking Fund Factor	Capital Recovery Factor	Compound Amount Factor	Present Worth Factor	Gradient Uniform Series	Gradient Present Worth	
	Find F Given P	Find P Given F	Find A Given F	Find A Given P	Find F Given A	Find P Given A	Find A Given G	Find P Given G	
n	F/P	P/F	A/F	A/P	F/A	P/A	A/G	P/G	n
1	1.120	.8929	1.0000	1.1200	1.000	0.893	0	0	1
2	1.254	.7972	.4717	.5917	2.120	1.690	0.472	0.797	2
3	1.405	.7118	.2963	.4163	3.374	2.402	0.925	2.221	3
4	1.574	.6355	.2092	.3292	4.779	3.037	1.359	4.127	4
5	1.762	.5674	.1574	.2774	6.353	3.605	1.775	6.397	5
6	1.974	.5066	.1232	.2432	8.115	4.111	2.172	8.930	6
7	2.211	.4523	.0991	.2191	10.089	4.564	2.551	11.644	7
8	2.476	.4039	.0813	.2013	12.300	4.968	2.913	14.471	8
9	2.773	.3606	.0677	.1877	14.776	5.328	3.257	17.356	9
10	3.106	.3220	.0570	.1770	17.549	5.650	3.585	20.254	10
11	3.479	.2875	.0484	.1684	20.655	5.938	3.895	23.129	11
12	3.896	.2567	.0414	.1614	24.133	6.194	4.190	25.952	12
13	4.363	.2292	.0357	.1557	28.029	6.424	4.468	28.702	13
14	4.887	.2046	.0309	.1509	32.393	6.628	4.732	31.362	14
15	5.474	.1827	.0268	.1468	37.280	6.811	4.980	33.920	15
16	6.130	.1631	.0234	.1434	42.753	6.974	5.215	36.367	16
17	6.866	.1456	.0205	.1405	48.884	7.120	5.435	38.697	17
18	7.690	.1300	.0179	.1379	55.750	7.250	5.643	40.908	18
19	8.613	.1161	.0158	.1358	63.440	7.366	5.838	42.998	19
20	9.646	.1037	.0139	.1339	72.052	7.469	6.020	44.968	20
21	10.804	.0926	.0122	.1322	81.699	7.562	6.191	46.819	21
22	12.100	.0826	.0108	.1308	92.503	7.645	6.351	48.554	22
23	13.552	.0738	.00956	.1296	104.603	7.718	6.501	50.178	24
24	15.179	.0659	.00846	.1285	118.155	7.784	6.641	51.693	24
25	17.000	.0588	.00750	.1275	133.334	7.843	6.771	53.105	25
26	19.040	.0525	.00665	.1267	150.334	7.896	6.892	54.418	26
27	21.325	.0469	.00590	.1259	169.374	7.943	7.005	55.637	27
28	23.884	.0419	.00524	.1252	190.699	7.984	7.110	56.767	28
29	26.750	.0374	.00466	.1247	214.583	8.022	7.207	57.814	29
30	29.960	.0334	.00414	.1241	241.333	8.055	7.297	58.782	30
31	33.555	.0298	.00369	.1237	271.293	8.085	7.381	59.676	31
32	37.582	.0266	.00328	.1233	304.848	8.112	7.459	60.501	32
33	42.092	.0238	.00292	.1229	342.429	8.135	7.530	61.261	33
34	47.143	.0212	.00260	.1226	384.521	8.157	7.596	61.961	34
35	52.800	.0189	.00232	.1223	431.663	8.176	7.658	62.605	35
40	93.051	.0107	.00130	.1213	767.091	8.244	7.899	65.116	40
45	163.988	.00610	.00074	.1207	1 358.2	8.283	8.057	66.734	45
50	289.002	.00346	.00042	.1204	2 400.0	8.304	8.160	67.762	50
55	509.321	.00196	.00024	.1202	4 236.0	8.317	8.225	68.408	55
60	897.597	.00111	.00013	.1201	7 471.6	8.324	8.266	68.810	60
65	1 581.9	.00063	.00008	.1201	13 173.9	8.328	8.292	69.058	65
70	2 787.8	.00036	.00004	.1200	23 223.3	8.330	8.308	69.210	70
75	4 913.1	.00020	.00002	.1200	40 933.8	8.332	8.318	69.303	75
80	8 658.5	.00012	.00001	.1200	72 145.7	8.332	8.324	69.359	80
85	15 259.2	.00007	.00001	.1200	127 151.7	8.333	8.328	69.393	85
90	26 891.9	.00004		.1200	224 091.1	8.333	8.330	69.414	90
95	47 392.8	.00002		.1200	394 931.4	8.333	8.331	69.426	95
100	83 522.3	.00001		.1200	696 010.5	8.333	8.332	69.434	100

Compound interest factors

18%									18%
	Single Payment		Uniform Payment Series				Uniform Gradient		
	Compound Amount Factor	Present Worth Factor	Sinking Fund Factor	Capital Recovery Factor	Compound Amount Factor	Present Worth Factor	Gradient Uniform Series	Gradient Present Worth	
n	Find F Given P F/P	Find P Given F P/F	Find A Given F A/F	Find A Given P A/P	Find F Given A F/A	Find P Given A P/A	Find A Given G A/G	Find P Given G P/G	n
1	1.180	.8475	1.0000	1.1800	1.000	0.847	0	0	1
2	1.392	.7182	.4587	.6387	2.180	1.566	0.459	0.718	2
3	1.643	.6086	.2799	.4599	3.572	2.174	0.890	1.935	3
4	1.939	.5158	.1917	.3717	5.215	2.690	1.295	3.483	4
5	2.288	.4371	.1398	.3198	7.154	3.127	1.673	5.231	5
6	2.700	.3704	.1059	.2859	9.442	3.498	2.025	7.083	6
7	3.185	.3139	.0824	.2624	12.142	3.812	2.353	8.967	7
8	3.759	.2660	.0652	.2452	15.327	4.078	2.656	10.829	8
9	4.435	.2255	.0524	.2324	19.086	4.303	2.936	12.633	9
10	5.234	.1911	.0425	.2225	23.521	4.494	3.194	14.352	10
11	6.176	.1619	.0348	.2148	28.755	4.656	3.430	15.972	11
12	7.288	.1372	.0286	.2086	34.931	4.793	3.647	17.481	12
13	8.599	.1163	.0237	.2037	42.219	4.910	3.845	18.877	13
14	10.147	.0985	.0197	.1997	50.818	5.008	4.025	20.158	14
15	11.974	.0835	.0164	.1964	60.965	5.092	4.189	21.327	15
16	14.129	.0708	.0137	.1937	72.939	5.162	4.337	22.389	16
17	16.672	.0600	.0115	.1915	87.068	5.222	4.471	23.348	17
18	19.673	.0508	.00964	.1896	103.740	5.273	4.592	24.212	18
19	23.214	.0431	.00810	.1881	123.413	5.316	4.700	24.988	19
20	27.393	.0365	.00682	.1868	146.628	5.353	4.798	25.681	20
21	32.324	.0309	.00575	.1857	174.021	5.384	4.885	26.330	21
22	38.142	.0262	.00485	.1848	206.345	5.410	4.963	26.851	22
23	45.008	.0222	.00409	.1841	244.487	5.432	5.033	27.339	24
24	53.109	.0188	.00345	.1835	289.494	5.451	5.095	27.772	24
25	62.669	.0160	.00292	.1829	342.603	5.467	5.150	28.155	25
26	73.949	.0135	.00247	.1825	405.272	5.480	5.199	28.494	26
27	87.260	.0115	.00209	.1821	479.221	5.492	5.243	28.791	27
28	102.966	.00971	.00177	.1818	566.480	5.502	5.281	29.054	28
29	121.500	.00823	.00149	.1815	669.447	5.510	5.315	29.284	29
30	143.370	.00697	.00126	.1813	790.947	5.517	5.345	29.486	30
31	169.177	.00591	.00107	.1811	934.317	5.523	5.371	29.664	31
32	199.629	.00501	.00091	.1809	1 103.5	5.528	5.394	29.819	32
33	235.562	.00425	.00077	.1808	1 303.1	5.532	5.415	29.955	33
34	277.963	.00360	.00065	.1806	1 538.7	5.536	5.433	30.074	34
35	327.997	.00305	.00055	.1806	1 816.6	5.539	5.449	30.177	35
40	750.377	.00133	.00024	.1802	4 163.2	5.548	5.502	30.527	40
45	1 716.7	.00058	.00010	.1801	9 531.6	5.552	5.529	30.701	45
50	3 927.3	.00025	.00005	.1800	21 813.0	5.554	5.543	30.786	50
55	8 984.8	.00011	.00002	.1800	49 910.1	5.555	5.549	30.827	55
60	20 555.1	.00005	.00001	.1800	114 189.4	5.555	5.553	30.846	60
65	47 025.1	.00002		.1800	261 244.7	5.555	5.554	30.856	65
70	107 581.9	.00001		.1800	597 671.7	5.556	5.555	30.860	70
75	46 122.1				1 367 339.2	5.556	5.555	30.862	75
100	15 424 131.9				85 689 616.2	5.556	5.555	30.864	100